HEARST MARINE BOOKS
TRAILERBOAT GUIDE

HEARST MARINE BOOKS TRAILERBOAT GUIDE

Joe Skorupa

Illustrations by Ron Carboni

Roy Attaway
Contributing Photographer

HEARST MARINE BOOKS
New York

Copyright © 1993 by Joe Skorupa

All rights reserved. No part of this book may be reproduced or utilized in any form or by any means, electronic or mechanical, including photocopying, recording, or by any information storage or retrieval system, without permission in writing from the Publisher. Inquiries should be addressed to Permissions Department, William Morrow and Company, Inc., 1350 Avenue of the Americas, New York, New York 10019.

It is the policy of William Morrow and Company, Inc., and its imprints and affiliates, recognizing the importance of preserving what has been written, to print the books we publish on acid-free paper, and we exert our best efforts to that end.

Library of Congress Cataloging-in-Publication Data

Skorupa, Joe.
    Hearst Marine Books trailerboat guide / by Joe Skorupa.
      p.   cm.
    Includes index.
    ISBN 0-688-12338-4
    1. Boat trailers. 2. Boats and boating. I. Title. II. Title: Trailerboat guide.
TL297.2.S57 1993
688.6—dc20   93-958
                                                                                     CIP

Printed in the United States of America

First Edition

1 2 3 4 5 6 7 8 9 10

Edited by Michael Mouland

BOOK DESIGN BY GIORGETTA BELL MCREE

CONTENTS

INTRODUCTION
ix

CHAPTER **1**
THE TRAILERBOAT
1

CHAPTER **2**
THE TRAILER
13

CHAPTER **3**
THE HITCH
25

CHAPTER **4**
THE TOW VEHICLE
35

CHAPTER **5**
ON THE ROAD
45

CHAPTER **6**
ON THE WATER
57

CHAPTER **7**
MAINTENANCE
71

CHAPTER **8**
ACCESSORIES
83

GLOSSARY OF BOATING TERMS
93

INDEX
101

INTRODUCTION

Boating adventure is in the eye of the beholder. A day of carving glass on water skis can be just as fun as a scenic cruise or a light-tackle hunt for fish. At least this is true for me. As a resident of northern New Jersey, I've had many memorable boating experiences on nearby Greenwood Lake—as well as Lake Hopatcong, the Hudson River, the Delaware River, and my recent passion, the Atlantic Ocean.

The wide range of available boating options has forced me into a difficult position. When the weekend rolls around and it's time to hitch up my boat, where will I launch—on the lake or on the river? Will I go fishing or cruising? Will I stay close to home or venture farther afield? How much farther afield? It's an interesting question.

As with most trailerboaters, the bulk of my experience is in modest-sized boats, and I've never thought this was a limiting factor. If anything, I've considered it something of a challenge. I've driven one twenty-foot boat up the Mississippi River from New Orleans to Minneapolis, and a twenty-two-foot boat from Miami to New York on the Intracoastal Waterway. I've towed a twenty-six-foot cruiser eighteen hundred miles from Cadillac, Michigan, to Key West, Florida, and crossed the Gulf Stream from Florida to the Bahamas in a twenty-five-foot center console.

As you can tell from these adventures, every now and then I get the urge to travel to exotic waters, and I don't let my trailerboat hold me back. Neither should you. Equipped with the proper knowledge to operate your rig—boat, motor, trailer, and tow vehicle—you can safely and efficiently go boating anywhere you choose.

The *Trailerboat Guide* presents this body of knowledge in a compact, easy-to-read, well-illustrated manner. It's an essential resource for beginners who've recently purchased their first boat. Equally, it's a vital manual for veterans who need to be updated on the latest developments in equipment, regulations, and the ever-changing standards for operation and safety. It covers the basics of boats, motors, trailers, and tow vehicles. It offers helpful tips on safe road travel, problem-free launching, and money-saving maintenance. It also presents up-to-date information on the latest developments in manufacturer-installed tow packages, trailer hitches, aftermarket accessories, and security devices that foil the growing number of boat bandits.

There are approximately six million trailerboaters nationwide, by far the largest segment in the boating market. Yet strangely, this huge bloc of boaters is frequently overlooked by the authors of boating books, guides, and manuals. This neglect results not only in less quantity of useful material, but in less quality, too. In fact, a great deal of what's available is often filled with out-of-date information that's of little use to trailerboaters. The *Trailerboat Guide* attempts to correct this situation.

So read the book and use it confidently to pursue your personal brand of boating. There's no guarantee that your boating experiences will be trouble free, but there's every certainty that they will be adventurous. Remember, trailerboaters aren't only kings of the waterway, they're also kings of the road, and only as limited as their dreams.

HEARST MARINE BOOKS
TRAILERBOAT GUIDE

CHAPTER 1

THE TRAILERBOAT

Wouldn't it be great if you spent all your boating time actually boating? Forget about pretrip preparation, long-distance travel, and vital maintenance. Just beam over to the lake and roar off into a bright blue paradise. Well, it doesn't happen that way. A large portion of every boater's time is spent on a wide variety of off-the-water activities. These are the fundamentals of boating and they'll be covered in great detail in the following chapters.

But trailerboaters are boaters first and everything else second. So, let's start by focusing on the love of a boater's life—the boat. Let's especially look at its design characteristics and how they affect trailering.

Down at the launch ramp no one has a problem defining a trailerboat. On a typical Saturday morning you see fiberglass and aluminum runabouts and fishing boats. You see cuddy and aft-cabin cruisers. You see deck boats, pontoon boats, water ski boats, and sport boats. You see center consoles, walk-around cabins, johnboats, walleye boats, flats boats, and bass boats. You see sailing skiffs and day sailers. And increasingly, you see jet-powered fun boats and personal watercraft.

The problem in defining a trailerboat arises not with boaters, but with boating books. Most of them typically use a definition that's so broad it includes everything from canoes to thirty-five-foot yachts hitched to an eighteen-wheeler.

Let's be clear about our definition. A *trailerboat* is a marine craft that's generally no more than twenty-six feet long and eight feet six inches wide. Combined with a trailer, the majority of these towing rigs generally weigh less than five thousand pounds.

Why limit the definition to these dimensions and specifications? Boats of this size are narrow enough to be street legal without a special permit and light enough to be towed by family-type vehicles. Consequently, they form the heart of the

HEARST MARINE BOOKS TRAILERBOAT GUIDE

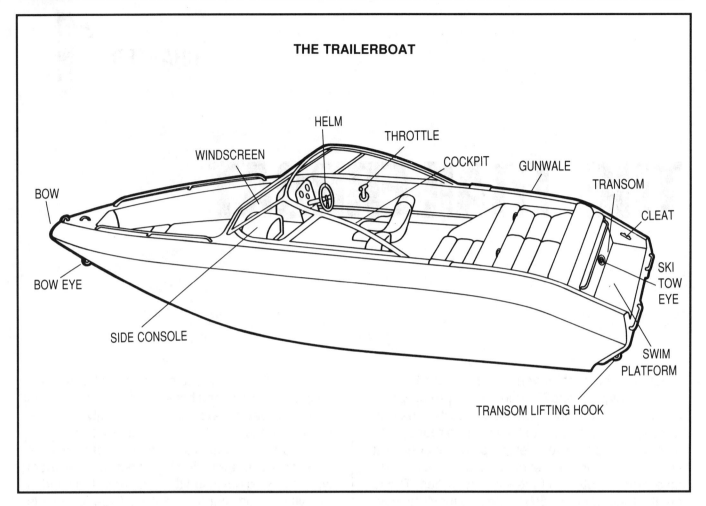

THE TRAILERBOAT

trailerboat market, and they're the primary focus of this book.

However, there are a few worthy exceptions that shouldn't be neglected. These include classic wooden boats, rubber inflatables, canoes, kayaks, and rafts, plus powerboats and sailboats that are longer, wider, and heavier than the norm. Where applicable, the needs of these special trailerboats will be addressed.

There are approximately six million trailerboats of all types and sizes. However, the typical trailerboat is best described as being twenty-six feet long or less, eight feet six inches wide or less, and five thousand pounds or less. Maximum towing weight ratings and highway restrictions are defining factors.

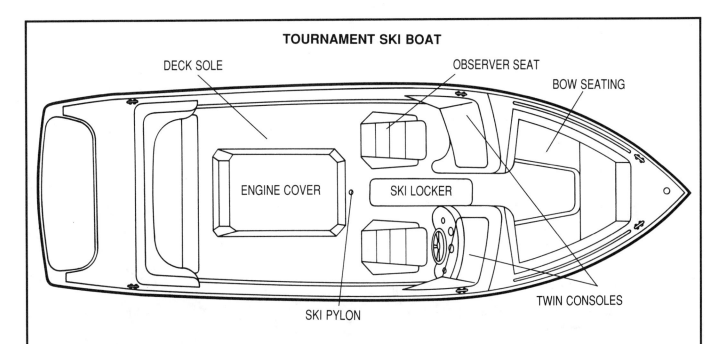

An inboard engine and a flat bottom give a dedicated ski boat an ideal wake for competition and training.

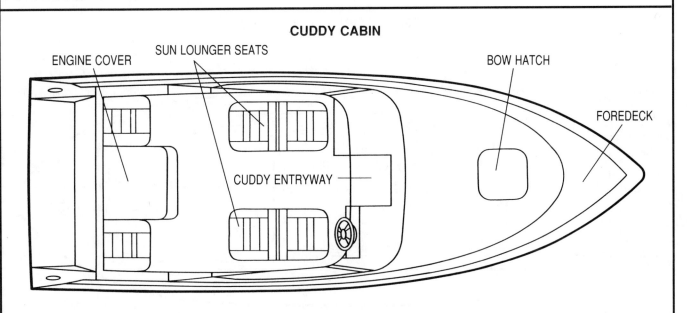

Cuddies range from enclosed foredecks on simple runabout hulls to overnight cruisers.

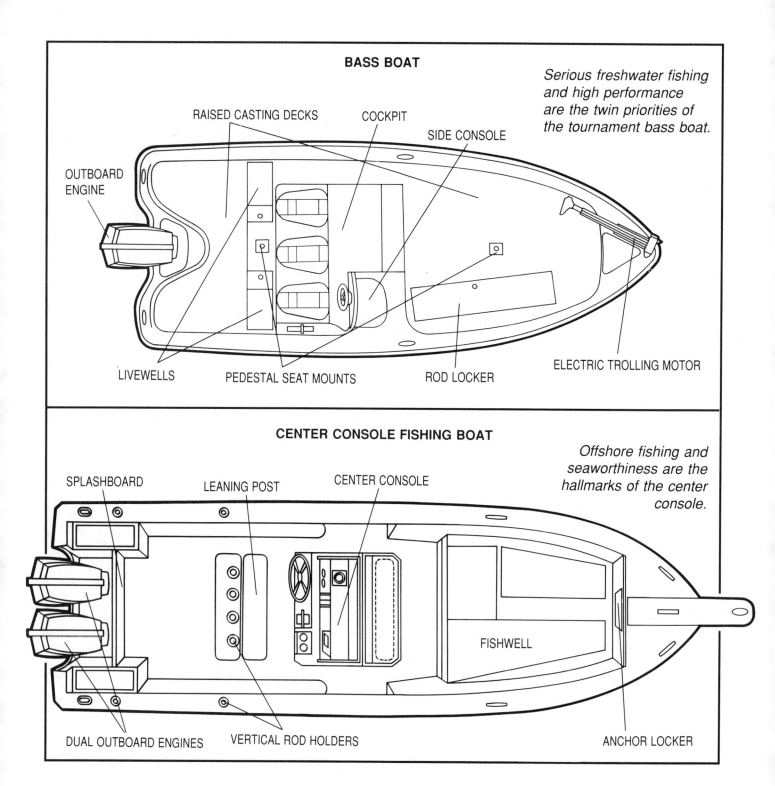

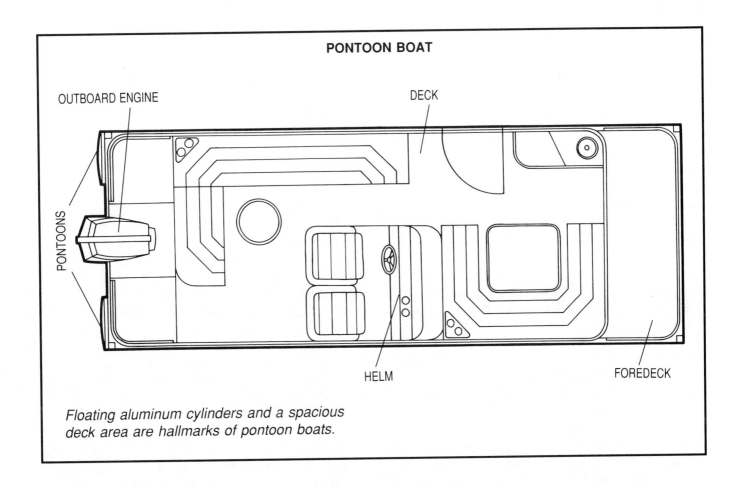

Floating aluminum cylinders and a spacious deck area are hallmarks of pontoon boats.

THE BASIC HULL

The component that most directly affects overall trailering and boat handling is the *hull*—the structural body of the boat that comes into contact with the water. It's also the part of the boat that comes into contact with the bed or cradle of the trailer.

The first things you need to know about the hull, as far as trailering is concerned, are its primary specifications: length, beam, height, and weight.

The *overall length* of the boat's hull, especially for trailering, is measured along the centerline from the *bow* to the *transom*. Such exterior components as the bow pulpit, swim platform, and outboard engine bracket are important for docking, storage, and performance, but don't necessarily affect the way the hull rests on the bed of the trailer. Until recently most of these components were add-on options, but today they're often integrated into the hull's basic mold.

Unfortunately, this situation has resulted in some confusion about how overall length is listed in manufacturer literature. Some manufacturers include the integrated components in their measurements. Others don't. Since there's no industry standard, the best advice is to read your owner's manual carefully to determine how the figure is obtained.

There's rarely any confusion about the boat's *beam* or width. It's always measured at the hull's widest point. On federal highways and on most state and local roads, the maximum beam allowable on the road is eight feet six inches. This limitation ensures that the vast majority of boats manufactured are narrower than this dimension and are therefore street legal.

The figure given for weight, however, isn't so straightforward. Many manufacturers use the term *dry weight* in their literature, which technically refers to the weight of the boat prior to filling the fuel and freshwater tanks. Since most boaters drive on the road with empty tanks, this seems like a logical standard. Except for one thing: It often refers to the weight of the boat prior to the installation of all options, equipment, gear, and most significantly, the engine. Again, the best advice is to read your manufacturer literature—carefully.

The final specification to note is height. Although most trailer boaters rarely worry about this dimension, overhead obstructions can make for unpleasant encounters. So, after mounting your boat on the trailer, measure the distance between the ground and the bottom of the hull. Then add this dimension to the height figure listed in the factory literature. This is your *trailering height*.

Naturally, owners of boats with aluminum towers, flying bridges, raised helms, and fixed masts will be more concerned with trailering height than the average boater. One reassuring note is that federal highways are generally built to accommodate tractor trailers and buses up to a maximum of thirteen feet six inches. If your trailering height is a foot or more below this benchmark, you won't have much to worry about on the interstates, but it's still wise to pay careful attention to posted signs. Once you get off the superhighways, the best advice is to pay even closer attention. As a matter of fact, this is good advice to follow at all times while trailering.

HULL SHAPES

To a trailerboater, the hull's bottom shape is as important on land, where it rests on the bed of the trailer, as it is on water.

Historically, trailerboat hulls have been variations on three basic themes—flat, round, and Vee. Of these, the most common are V-shaped *planing hulls* characterized by hard *chines* and the ability to lift partially out of the water at running speeds. These planing hulls fall into two categories: tri-hull (or cathedral) and classic wedge-shaped Vees.

As the name *tri-hull* implies, boats with this design are loosely characterized by three V-shaped, side-by-side bottom components on the hull. Sometimes called *cathedral hulls,* they perform with efficiency and have outstanding side-to-side stability. However, they tend to pound in rough water and give a relatively wet ride. For these reasons, tri-hull boats have been phased out over the years, although many old models still ply the waterways.

Tri-hulls have been superseded in modern times by more wedgelike *V-hulls.* These hulls are differentiated from each other by the angle of the Vee measured at the transom. This angle is stated in degrees of *deadrise*. The industry has no rigid standard to categorize V-hulls, but the following classes may be used as a guide: 1. *Deep-V* (18° to 24°), used primarily on offshore boats; 2. *Mod-V* (12° to 17°), which can be used on either near-shore saltwater or big-lake freshwater boats, and 3. *Flat-V* (less than 12°), which is used primarily on lake or river boats.

THE TRAILERBOAT

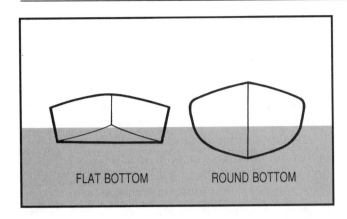

The two oldest hull designs are flat bottoms, which are found on small dinghies and rowboats, and round bottoms, which can be either full displacement or semidisplacement hulls.

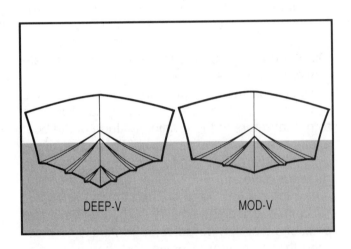

Both deep-V and mod-V hulls are planing hulls that use bottom strakes to help reduce the wetted surface and improve handling.

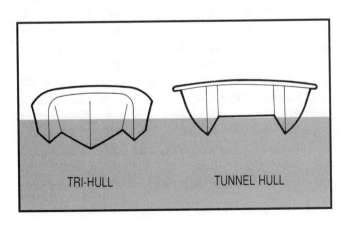

The tri-hull (or cathedral hull) is characterized by having three hull points in contact with the water. The tunnel hull rides on two points called sponsons.

The advantage of the V-hull design is its sharp entry into the water, which makes for exceptional straight-line tracking and holding tight in turns. Equally important, its sharp entry enables the boat to slice into the chop instead of pounding down hard on each wave.

Few boaters today encounter the simple but rough-riding *flat hull*. Exceptions are owners of johnboats, dinghies, small sailboats, and aluminum rowboats or fishing boats. Equally rare are *round-bottom* displacement or *semidisplacement* hulls. These designs are generally confined to vintage wooden boats or sailboats.

Pontoon boats and tunnel hulls are more common. *Pontoon boats* are characterized by twin airtight, semidisplacement hulls connected above the waterline by a platform deck. Inexpensive, low-powered pontoon boats are traditionally made with welded-aluminum pontoons. Some builders today, however, are starting to use fiberglass pontoons that incorporate some planing characteristics to improve efficiency.

Tunnel-hull boats use the twin-hull design to achieve maximum hair-straightening speed. The twin hulls are designed to trap air beneath the boat and to lift it out of the water for minimum drag. When the boat is running at wide-open throttle, the hulls lightly kiss the water's surface. In addition to speed, tunnel boats have terrific handling characteristics.

A typical tunnel boat has a raised center section flanked by *sponsons* on either side. The sponsons can be either catamaran style, which are symmetrical, or *dihedral*, which are half-Vees. An interesting tunnel boat variation is the *mod-VP*, which has a V-bottom flanked by two sponsons. This design offers the best of both worlds—riding comfort and air entrapment.

BASIC POWER

Like the hull shape, the boat's power source is as important on land as it is on water. Each type of marine power, either engine-driven or wind-driven, affects how a boat rests on the bed of a trailer.

The most common type of power found in boating is supplied by an *outboard* motor, which is bolted onto the back of the boat's transom. Outboards are typically internal-combustion *two-cycle* engines, which means they have a power stroke every other cycle of the piston. An outboard can be portable if the motor is light enough to carry, however, most outboards are too heavy to carry easily, and are permanently fixed to the transom. Outboard power ranges from as low as 2 horsepower to 300 horsepower.

Most midsize and large outboards are equipped with power trim and steering systems. *Trim* is a term that refers to the running attitude of the boat as it's affected by the position of the motor's drive unit. Outboards equipped with power trim can raise the drive unit or lower it by pressing a control button on the throttle or on the steering wheel.

Some midsize and smaller outboards have trim positions that can be adjusted manually by muscling the engine up or down and then fixing it in place. Most outboards in this power range, however, have no trim capability. The majority of these are *tiller-handle* motors, which require the boater to operate the engine by holding on to a control arm and then manually turning the engine from side to side while under power. Both tiller-handle and steering-equipped outboards can be tilted up and out of the way when the boat is loaded onto a trailer.

Two rarer kinds of outboards sometimes found on trailerable boats are the four-cycle motor and the inside-mounted outboard. The *four-cycle* out-

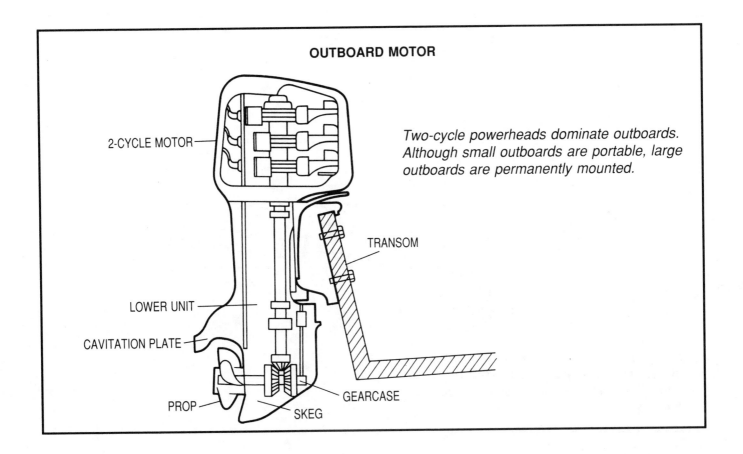

Two-cycle powerheads dominate outboards. Although small outboards are portable, large outboards are permanently mounted.

board is an internal-combustion engine that has a power stroke once every four cycles of the piston. The extra two cycles and the addition of valve assemblies and other components enable the engine to reduce emissions and improve fuel efficiency. Four-cycle outboards are frequently used on sailboats as auxiliary motors.

The *inside outboard* made its debut recently in midsize runabouts, and it's exactly what it sounds like—a two-cycle outboard that's fully enclosed by an engine compartment. The advantages of this type of motor are quiet operation and a favorable power-to-weight ratio compared to inboard engines.

True *inboard engines* are located inside the hull of the boat and are positioned amidships or somewhat aft. They are typically four-cycle motors that consist of automotive engine blocks that have been *marinized*—fixed with components and systems, such as raw-water cooling systems and flame arrestors, that enable them to cope with the harsh marine environment.

A second type of inboard engine is the *diesel,* which comes from the automotive world and also requires marinization. Diesels are four-cycle internal-combustion engines that generate tremendous compression to ignite small quantities of low-combustion diesel fuel. Compared to gas en-

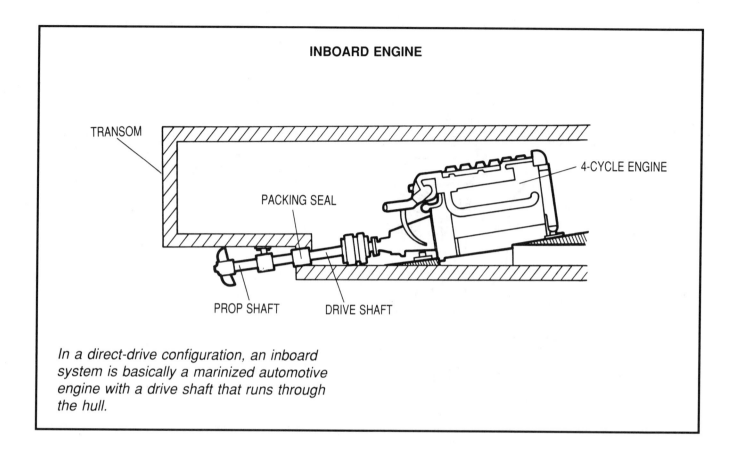

In a direct-drive configuration, an inboard system is basically a marinized automotive engine with a drive shaft that runs through the hull.

gines, diesels are very fuel-efficient, long-lived, and inexpensive to maintain. However, they're also relatively heavy and pricey—at least twice the cost of comparably powered gas engines. For this reason diesels are more common in yacht-sized vessels than in trailerboats.

An inboard engine runs a drive shaft through the bottom of the hull in two different ways: *direct drive,* which calls for the engine to be tilted on its mountings so that the shaft can run in a straight line through the bottom of hull; and *V-drive,* which splits the forward facing shaft to form a V-angle before running the prop shaft back through the hull. Unlike outboards, the prop shafts on inboard engines cannot tilt out of the way during trailering.

An early name for the *stern-drive engine* was *I/O,* which stood for *inboard/outboard.* This is an apt name because stern drives are a unique combination of features found on both inboard and outboard engines. The typical stern drive has a four-cycle, marinized automotive engine mounted inside the boat just forward of the transom. The vast majority use gasoline engines. However, there are a few diesel stern drives on U.S. waterways, although they're more popular internationally.

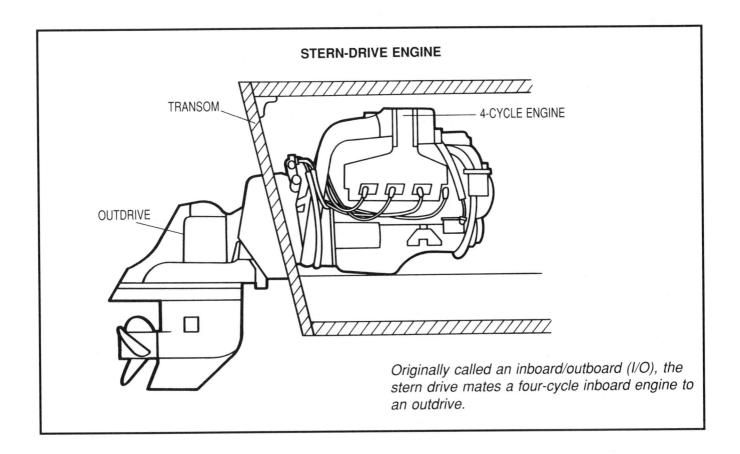

Originally called an inboard/outboard (I/O), the stern drive mates a four-cycle inboard engine to an outdrive.

What makes the stern drive interesting is that instead of running the drive shaft at an angle through the bottom of the hull, it runs the shaft horizontally through a cutout in the transom. Here, the shaft connects to an *outdrive* unit that runs vertically down into the water. It's a complicated piece of engineering that requires splitting the drive shaft twice to form a Z-shape. After the ubiquitous outboard, the stern drive is the second most common engine in trailerboating. Modern stern drives are generally equipped with power trim, and their lower units can be tilted out of the way during trailering.

Power components on sailboats have a direct effect on trailering, too. The two with the most impact are the *mast* and the *keel*. The mast can be either stepped, which means that it can be taken down, or fixed, which means that it's permanently installed. The keel can be either retractable, which means that it can slide up and out of the way, or deep, which means that it's fixed in place and the bulb is filled with weight to act as ballast.

The best of all possible worlds for a trailerboater is to own a sailboat with a retractable keel and a stepped mast. Boats like these, which include small skiffs and catamarans, are easy to launch and retrieve. They also ride low on the trailer and present a problem-free profile.

Deep-keel and fixed-mast sailboats make trailering a bit more difficult. To accommodate the deep keel the trailer must be fitted with a tall cradle. This makes using a launch ramp virtually impossible. Boats of this type are generally launched by using a hoist or crane. Owners of sailboats like these generally spend as little time as possible on the road; however, such boats can be transported over land when necessary.

Actually, just about any boat can become a road warrior if it's matched to the right trailer.

CHAPTER 2

THE TRAILER

Today, you can go to a marine dealer and pick up a dream boat right off the showroom floor. This wasn't always the case. Until recently, virtually every element of a trailerboat was sold separately and then tacked on to the sticker price. In many instances, the price doubled before you could get the boat out the door. This situation didn't thrill many buyers, and boat builders eventually streamlined the process.

Now, just like automobiles, boats are prerigged by builders (as opposed to dealers) and sold fully equipped. This is especially true in the trailerboat class, where most models come with a designated motor, trailer, and a long list of standard equipment. Most have only a few options available. Many have none.

These *package boats* have had a major impact on how boats are matched to trailers. On the one hand, they've shifted the trailer-buying decision from boaters and dealers to boat builders, who are in a perfect position to select trailers for their boats. Builders know all there is to know about each boat's specifications, method of construction, and support structure. They also have a vested interest in making sure the selection process is done right, because in a prerigged situation the trailer will ultimately reflect on overall buyer satisfaction.

But there's a downside, too. Since the selection is now out of the buyer's hands, most boaters have a reduced incentive to become familiar with trailer basics, which is unfortunate.

There's a lot you need to know about your trailer, including tips and techniques for preventative maintenance, repairs, safety on the road, efficiency on the launch ramp, and other hard-to-

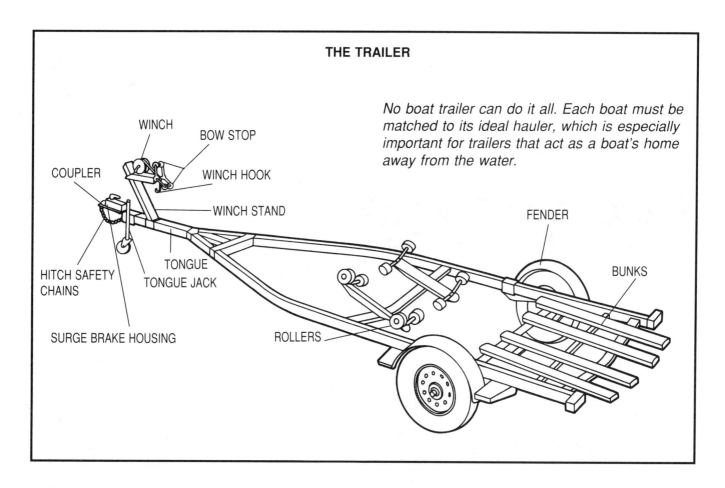

THE TRAILER

No boat trailer can do it all. Each boat must be matched to its ideal hauler, which is especially important for trailers that act as a boat's home away from the water.

come-by information. All of this proceeds from a solid grounding in the basic structure of the trailer itself, which is often a home away from home for much of a boat's life.

TYPES OF TRAILERS

A trailer, by definition, is a load-bearing vehicle that enables a boat to be towed by a car or a truck. It typically consists of an all-steel tube frame that's either painted (for freshwater) or galvanized (for saltwater). There are some exceptions, such as trailers made of relatively expensive aluminum and I-beam or C-beam construction, but these are relatively rare.

Beyond the basic frame, there are three types of common trailer designs: 1. the bunk trailer; 2. the roller trailer; and 3. the flatbed trailer. Each is satisfactory when properly matched to a boat.

The simple *bunk trailer* is the most common trailer in boating. It's suitable for a wide range of hulls, but especially for easy-to-launch boats that typically use deepwater ramps. The adjustable bunks, which are located at contact points and form the trailer cradle, are carpeted for cushioning and generally made of wood.

Bunks perform several important functions.

THE TRAILER

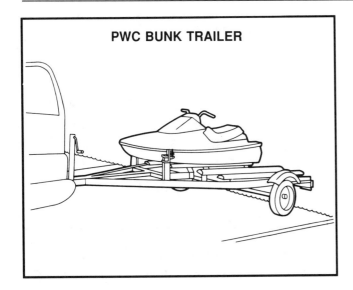

Versatile bunk trailers are used for small personal watercraft (PWC) as well as big cruisers.

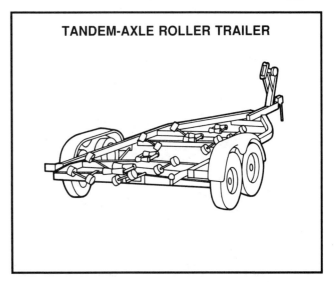

Big boats often require two axles for road support and rollers for launching ease.

First, they position the boat so that it balances properly over the axle. Second, they hold the boat securely in place by lining up with hull points that are internally supported. And third, they spread out the points of stress to prevent potential hull deformities, which can result from poor alignment.

Roller trailers use rubber or plastic *rollers* in place of bunks. The rollers are usually mounted on adjustable brackets or they can be fixed. The result is the smoothest and easiest method of moving the boat on and off the trailer. Roller trailers are especially well suited for difficult-to-launch boats and shallow-water ramps.

Compared to bunk trailers, roller trailers are equipped with extra hardware—often a dozen or more roller assemblies and a handful of brackets. These add cost and require increased maintenance. Many boaters believe the benefits of low-friction launching and retrieval outweigh these disadvantages; however, the advent of modern, well-designed, drive-on bunk trailers have greatly reduced the need for rollers.

Bunk and roller features aren't mutually exclusive. Many trailers combine the best of both worlds. *Keel rollers,* for example, are commonly found on bunk trailers, especially for midsize and large boats. The keel rollers help ease the boat on and off the trailer, and provide support at a pivotal point in the cradle. Conversely, keel pads, rear bunk assemblies, and forward bunk pads appear

HEARST MARINE BOOKS TRAILERBOAT GUIDE

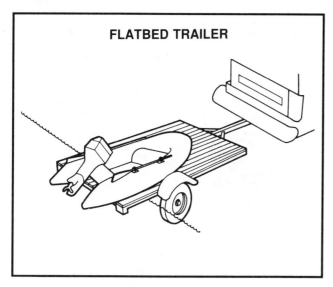

Boaters who use flatbeds for inflatables and other light craft can also use them for nonmarine hauling duties.

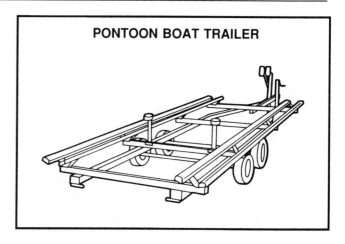

The long bunks of the pontoon boat trailer require tandem axles to improve road handling.

on roller trailers. The additional bunks help capture the boat during retrieval and improve hull support while on the road.

To the typical boater, the *flatbed trailer* refers to a relatively small trailer with a simple bed made of planks or plywood. These carryall workhorses have many uses, but they're generally limited to light, specialized craft. Although flatbeds can be equipped with fixed or temporary cradles to hold large V-bottom, semidisplacement, or deep-keel boats, rigs like these require a hoist for launching, and therefore see infrequent service by the vast majority of trailer boaters.

The beauty of the flatbed design is that it's so simple and versatile that it works as well in the tractor-trailer class for large yachts as it does for owners of personal watercraft, flat-bottom aluminum boats, pontoon boats, canoes, kayaks, and inflatables.

A flatbed also works well for the boater who wants to use it for double duty: annual boat launching and retrieving (through the use of a removable cradle) plus utility work. For these boaters, it should be noted that bunk trailers can be converted for double duty, too. The conversion from a bunk to a flatbed involves removing the bunk brackets and then bolting down a sturdy wooden frame and bed.

Without bunks or rollers to hold a boat in place, flatbeds are commonly equipped with guide bars, tie-downs, support racks, and other add-on accessories. One feature that's sometimes used on

THE TRAILER

small flatbed trailers is a hinged tongue or hinged frame, which enables the bed to tilt like a dump truck. Although *tilt-bed trailers* aren't recommended for craft of substantial size or weight. They're basically used for boats small enough to be muscled around during launching and retrieval.

TIRES

On the highway, the vital trailer components and systems located at frame level or below are called upon to perform extremely rugged duty. These components include: the axle, tires, wheels, wheel bearings, suspension system, and brakes. Most are part of the vital undercarriage or platform, which is designed to give the trailer efficient and safe mobility.

Starting from the ground up, trailer tires come in three basic types: diagonal bias, belted bias, and radial.

Diagonal bias tires are characterized by reinforcing plies or layers of threaded fabric (either nylon or polyester) that crisscross through the tread area and sidewalls. *Belted bias* tires have similar construction, but feature additional plies of fabric

Diagonal bias and belted bias tires are simple and affordable, but they have been surpassed by durable automotive-type radial tires.

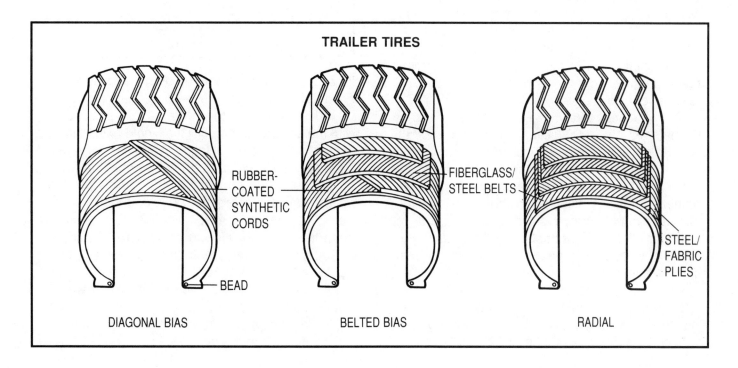

TRAILER TIRES — DIAGONAL BIAS, BELTED BIAS, RADIAL

running through the tread area. The development of radial tires, however, has made belted bias tires all but obsolete.

Radial tires, the most common automotive tires on the road today, use a completely different construction method. Instead of crisscrossing plies of fabric diagonally, radial tires have multiple plies of fabric with the cords or thread running at a 90-degree angle from the centerline. The line of the cords runs directly across the face of the tread area and radially through the sidewalls. For extra strength, a belt of steel wire or threaded fabric runs through the tread area.

The future clearly belongs to radial tires, and bias tires are being phased out for automotive use. Because of this situation, it's important to note that even if your tow vehicle and trailer have the same tire size, it's not wise to use them interchangeably.

The tire industry rates passenger and trailer tires differently for load limits. Trailer tires are built with greater sidewall strength to resist impact bruises and breaks. They're also designed to be used at high pressure ratings. If you must use passenger car tires on your trailer, check the load limit rating and reduce it by a minimum of 10 percent to give yourself a safety cushion.

Many trailers come with tires that are smaller in diameter than automotive tires. These small wheels position the axle and load closer to the ground for a low and stable center of gravity. But small tires turn at high rpm and sink deeper into potholes, which translates into shorter running life. The best recommendation is to confine trailers with small tires to light boats and short hauls. For heavy rigs and long hauls, it's best to go with smooth-riding large tires.

The most important things a boater needs to know about tires and their use on trailers are their size and maximum load capacity. By industry convention, this information should be provided by a

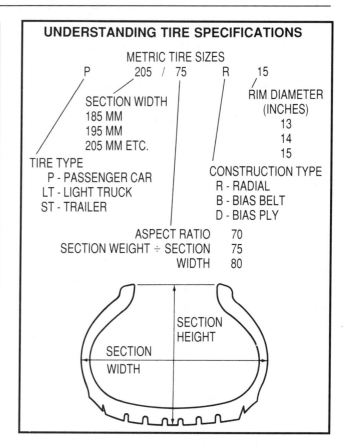

label placed on the front half of the left side of the trailer, either on the tongue or the main frame.

Here's how to use the information to select tires with the correct load rating: If a single-axle trailer's Gross Axle Weight Rating (GAWR) is 4,000 pounds, then each tire on the axle must be rated to at least 2,000 pounds (2,000 pounds per tire multiplied by 2 tires equals 4,000 pounds). If a tandem-axle trailer's GAWR is 4,000 pounds, then each tire must be rated to at least 1,000 pounds (1,000 pounds per tire multiplied times 4 tires equals 4,000 pounds.).

When in doubt about your trailer's exact weight (either empty or fully loaded), find a local truck scale and weigh it. A small fee may be charged, but it's well worth the cost.

THE TRAILER

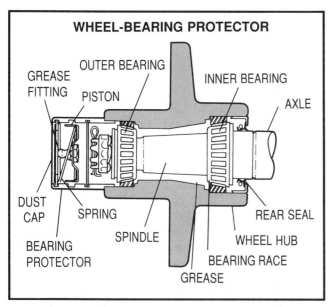

Sooner or later all boat trailers submerge their wheels, which means they should be equipped with bearing protectors to keep the wheel bearings packed with lubricant.

WHEEL BEARINGS

The components designed to allow the tires and wheels to roll freely are the *wheel bearings,* which are two rings of steel rollers located inside the wheel hubs. The wheel bearings rotate around a part of the axle called the *spindle.* For the bearings to work properly, the hubs need to be well packed with lubricant. Repacking the wheel bearings is a messy and laborious job.

Fortunately for boaters, *wheel-bearing protectors* eliminate the necessity to do frequent repacking. Essentially, wheel-bearing protectors fit over the hub and help keep grease in and water out by maintaining a positive internal pressure through the action of a steel spring and seal.

The problem of water intrusion is a serious one for wheel bearings, and it arises through the conditions of normal use. The hubs and bearings get hot during the trip to the launch ramp. Then, they're immediately immersed in cold water. The result is to draw water directly onto the bearing surfaces through the seals. Without the positive pressure of the bearing protectors, boaters would need to repack the wheel bearings every time they launched. With bearing protectors installed, all you need to do is replenish lubricant as needed through a grease fitting in the protector cap.

Bearing protectors are rapidly becoming a standard item in the trailer industry, and are increasingly found on quality trailers as standard equipment, especially in saltwater areas. Aftermarket models are available to fit most hubs, and they're well worth the added cost.

AXLES

Basically, there are two types of axles: *straight,* which is the most common, and *drop,* which features a center segment that's lowered up to four inches. Drop axles are typically used for tall boats to help lower the center of gravity.

One of the most important axle considerations for a boater is whether to buy a trailer with a single- or *tandem-axle* configuration. By distributing the boat's weight over two axles and suspension systems, a trailer can achieve improved on-road smoothness and stability, reduced tendency to wander, decreased sensitivity to tongue weight, and an increased margin of safety in case of a flat tire. To help accomplish this, a good tandem-axle trailer will be equipped with load equalizer bars that distribute the shock loads between the adjoining suspension systems and minimize the impact.

Tandem-axle trailers are recommended for boaters who carry heavy loads and haul long distances.

Although securely bolted in place, the axle or axles can be moved on the trailer frame, if necessary, to properly balance the load and achieve the ideal tongue weight.

BRAKES

The NMMA recommends that a trailer should be equipped with brakes if it has a minimum Gross Vehicle Weight Rating (GVWR) of 1,500 pounds. *GVWR* refers to the sum of the trailer's weight plus its maximum carrying capacity.

Trailer brakes must be able to operate automatically when the tow vehicle's brakes are applied, and in an emergency when the trailer becomes separated from the tow vehicle. Basically, there are two types of brakes common in trailerboating: hydraulic brakes and electronic brakes. Surge brakes, which are triggered by an actuator, are a type of hydraulic brake system.

Surge brakes are the most common type of brake system found on boat trailers. They work by sensing a sudden slowing of forward momentum or dipping of the tow vehicle's rear bumper, and then triggering the master brake cylinder. During normal running, a plunger or actuator remains in a neutral position. When forward momentum is slowed on a typical surge-brake system, the plunger or actuator automatically moves forward and trips a switch on the master brake cylinder. On some heavy-load trailers, surge brakes are activated by a telescoping action of the trailer tongue caused by the slowing of momentum.

In case the trailer becomes separated from the tow vehicle while running, the brakes are fitted

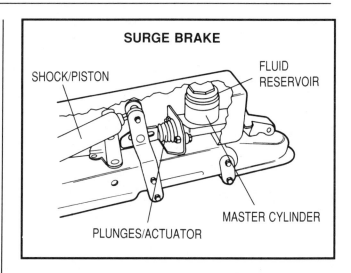

A sudden slowing of forward momentum activates the hydraulic surge brake.

with an emergency stop system. This consists of a short lanyard or cable that runs from the master brake cylinder to the receiver hitch on the tow vehicle. This lanyard is connected by hand before each trip. If the trailer becomes loose, it pulls a lever assembly forward and triggers the master brake cylinder.

In boating, hydraulic surge brakes have several advantages compared to *electronic brakes*, which are activated by pressing a button on the floor of the tow vehicle. Unfortunately, they have a tendency to short out after being immersed. While electronic brakes offer the driver a measure of increased control on the road, especially when just a light touch of braking is needed, surge brakes are easy to maintain, very durable in marine conditions, work well, and don't require special wiring. Mechanically simple drum-type brakes, as opposed to disc brakes, are standard throughout the trailerboat industry.

THE TRAILER

SUSPENSION SYSTEMS

To avoid punishing metal-to-metal connections, trailer frames are suspended from their axles by means of three different systems. These are leaf springs, coil springs, and torsion bars. Of the three, leaf-spring systems are the most common.

Like wheel bearing protectors and surge brakes, *leaf springs* are mechanically simple. The axle is suspended below the frame by resting on leaves or strips of steel, which flex when the wheels go over a bump. The leaves are connected to the frame by hangers and shackles. Boats as light as canoes, kayaks, inflatables, and single-man racing sailboats may only need a single leaf, while larger boats may use seven or more.

Coil springs, often employed in conjunction with shock absorbers in a system called coil-over shocks, are far more common in cars and trucks than they are in boat trailers. Like electronic trailer brakes, they work well but are relatively complex, vulnerable to corrosion, and harder to maintain.

Torsion bar systems are fairly new to boat trailers, although they've been around for years in other applications, especially RVs. Essentially, this maintenance-free, independently acting system consists of a hexagonal exterior axle made of tubular steel that encloses a three-sided solid-steel shaft. The solid shaft in turn is surrounded by three rubber inserts. At the ends of the axle, the wheels are mounted on hubs connected to spindles, which extend from the solid inner shaft.

When a wheel goes over a bump, the spindle transfers a twisting force to the solid shaft and the rubber inserts absorb it. It's an effective design that produces a soft, independently acting, rattle-free ride.

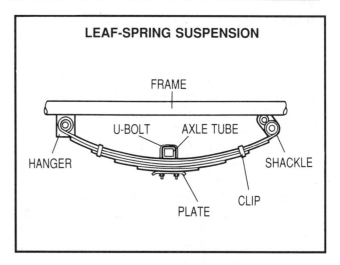

For light loads, a single steel leaf can provide adequate suspension. Heavier loads require multiple steel leaves.

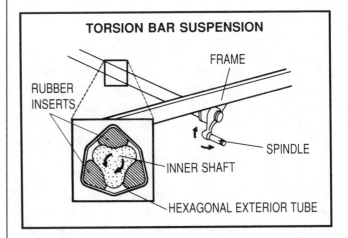

With a torsion bar, bumps in the road force the spindle to transfer a twisting motion to the steel inner axle, which is absorbed by rubber inserts.

LIGHTS

Both sides of the rear on a trailer are required to have combination reflector-type lights that perform the following functions: stoplight (red), taillight (red), turn signal (amber or red), and side marker (red). Another light is required for the license plate (white), and both sides of the trailer must have side-marker lights (red in rear and amber in front). Side and rear reflectors are also required.

Requirements for trailers wider than eighty inches have several additions: A tight cluster of three identification lamps (red) along the centerline on the rear; four clearance lamps with two on either side of the front (amber) and either side of the rear (red); and the stop lamps and turn signals must not be less than twelve square inches in area. Regardless of size, taillights must be set a minimum of fifteen inches above the ground and a maximum of seventy-two inches high for stop lamps and eighty-three inches high for turn signal lamps. Trailers greater than 30 feet in length, less than six feet in length, or less than thirty inches wide have fewer requirements.

Trailer lights are powered and activated by the tow vehicle's electrical system, which is connected to the trailer typically by means of a four-prong plug. Although there are exceptions, a color-coded wiring scheme is standard: brown wire for taillights, rear-marker lights, side-marker lights, and license light; yellow wire for left stop and turn lights; green wire for right stop and turn lights; and white wire for the ground. Plugs that have five, six, or seven prongs are sometimes found on trailers, especially for nonmarine use, but the four-prong plug is ubiquitous in boating.

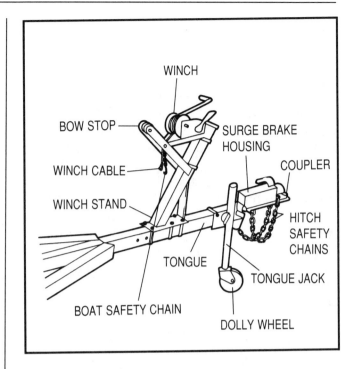

The tongue end of a trailer consists of components that secure the boat to the trailer and the trailer to the hitch.

FORWARD COMPONENTS

The forward part of the trailer is called the *tongue*, and several important components are located here. These include the winch, the winch stand, the bow stop, the bow-eye safety chain, and the tongue jack. Several other components are associated with the coupler, and these will be covered in Chapter 3.

The tongue itself can have several configurations. The hinged tongue of a tilt-bed trailer is one

THE TRAILER

example already covered. Two others are removable tongues and extension tongues.

A removable tongue enables a boater to fit a trailer into a space where length may be a problem, and provides a measure of security against theft. Without a tongue, a trailer is difficult to steal. Extension tongues are used to enable a trailer to back far into the water without submerging the tow vehicle. Deep-keel boats and shallow launch ramps may require the use of tongue extensions. Some trailers have a third tongue feature, a swing-away hinge, which has some space-saving value during storage. All three tongues (four if you include the hinged tongue) are relatively rare.

The strong pulling power of a *winch* is primarily used to help control the boat during retrieval and launching. It also enables a boater to snug a boat to the bow stop on the winch stand and solidly hold it in place for on-road support and security.

There are two kinds of common winches: manual hand-crank (one-speed and two-speed) and electric. The electric winch is the ultimate in no-effort retrieval, but most owners of small and mid-size boats get by with mechanically simple and affordable manual winches. For large boats, two-speed manual or electric winches are recommended. Electric winches are connected to the tow vehicle's battery by a special wiring harness.

For the winch to work properly, it should be mounted at the same height as the boat's bow eye or above it. The connection between the winch and the bow eye is handled by a steel hook and a length of woven synthetic strap or steel cable. The length of the strap or cable should be significantly greater than the length of the boat.

Although the winch stand or post is fixed firmly in place during use, its position is adjustable. Ideally, it should be placed so that the transom rests directly over supporting bunks or rollers. Bolted in place, the winch stand acts as an anchor for the boat during all phases of trailering.

To cushion the boat/trailer connection against the winch stand, rubber blocks or stops are used. This is called a *bow stop*. A short safety chain hangs on the bow stop and hooks to the bow eye. The chain acts as an emergency backup in case the winch line loosens or breaks.

The final component associated with the tongue is a jack. The *tongue jack* enables the boater to raise and lower the tongue. There are two types of tongue jacks: a drop-through jack, which simply moves up and down and often has a removable dolly wheel or steel foot; and a swivel jack, which pivots out of the way when not in use.

MATCHING YOUR BOAT TO A TRAILER

When the need arises to buy a trailer, there are no simple formulas, but here are several recommendations to follow:

- If your boating is done exclusively in freshwater, a painted finish is all you need. But if a part of your boating is done in saltwater, even a small part, a trailer with a galvanized finish is recommended.
- Bunk-type and roller-type trailers can be adapted to any kind of hull configuration, although rollers are unnecessary for boats that require a hoist for launching and retrieval. If you often launch at ramps under conditions that make driving on difficult, then roller-type trailers will provide an advantage.
- Keep in mind that it's important for the hull to be well supported along the chines, at the transom (especially for outboard-powered boats), along the keel, and at other points where there's internal structural support. The more points of contact there are between the boat and the

trailer, the better. This will spread out stress and help avoid hull deformities.
- If your trailer will be used as a dry dock for significant periods of time, bunk-type trailers have an edge because of their long points of contact. Roller-type trailers can work, too, if the rollers are large and numerous.
- Flatbed trailers can be adapted to any kind of hull, but they require cradles or racks for anything other than flat-bottom boats.
- If your boat is light enough to be muscled around, a tilt-bed trailer can be effective. However, most boats are too heavy to take advantage of these benefits.
- Trailers less than 80 inches wide are ideal for small and medium-size boats up to roughly 2,000 pounds. Wider trailers, up to 102 inches wide, can handle loads well beyond 10,000 pounds. Wide trailers are more stable on the road than narrow ones, and they're also more costly. It's important to note that the maximum width recommended for trailer loads is 102 inches (8.5 feet) in most states. Boats that are wider require special permits.
- To calculate the carrying capacity you need for your trailer, start with the dry weight of your boat and then add up all the extras. This will include the engine (if not included in the dry weight), anchor and chain, lines, fenders, Coast Guard–required safety equipment, water skis, fishing gear, spare parts, and other accessories. Also, estimate how much you may carry in water (8.3 pounds per gallon), fuel (6.2 pounds per gallon), and oil (7 pounds per gallon). And finally, figure the weight of an ice chest stocked with beverages and food. When this is tallied, it's a good idea to add another 20 percent for a safety margin, even if you have to move up a size. You don't want to skimp on an undersized trailer.
- If your boat weighs less than 2,500 pounds, a single-axle trailer will generally be adequate. If your boat is heavier than 2,500 pounds, give serious consideration to a tandem-axle rig for increased on-road stability. For loads in the 10,000-pound range, your best bet may be a triple-axle rig.
- A correctly balanced trailer carries most of its weight over the axle or axles and only a small percentage of weight forward on the tongue. Most hitches rate maximum tongue weight to be 10 percent of the trailer's GVWR, although most boaters can safely get by with a few percentage points less of GVWR.
- In the final analysis, the best trailer for your boat will be one that: 1. has enough support points to hold your boat in place without inflicting undue hull stress; 2. is stable at highway speeds; and 3. permits easy launching and retrieval.

CHAPTER 3

THE HITCH

Whether you're maneuvering around the launch ramp or driving down the road, the vital link between your tow vehicle and trailer is the hitch. When properly deployed, the hitch allows your trailer rig to work together smoothly and safely.

In general, the *hitch* is a fastening mechanism that connects a movable load carrier to a vehicle that pulls it. Specifically, it's a trailer mounting system that's fixed to a car or truck. It consists of a frame that's bolted to multiple points on the tow vehicle's undercarriage, a fixed ball-mount platform or receiver box, hooks for attaching safety chains, and a hitch ball.

Several other components are needed to complete the hitch connection. These are located on the trailer tongue and include a coupler, an electric wiring harness, safety chains, and a surge-brake cable. While they aren't technically part of the hitch, they're indispensable to the working link between the trailer and the tow vehicle.

HITCH CLASSIFICATIONS

Hitches, couplers, and balls are all weight-bearing components, and all use maximum capacity ratings based on the GVWR of the trailer. Additionally, hitches are rated for tongue weight. As noted in Chapter 2, boaters are best served by tongue weights ranging from 5 to 7 percent of the trailer's GVWR, with a maximum limit of 10 percent.

Of the two weight ratings for hitches, the more important one is based on the trailer's GVWR.

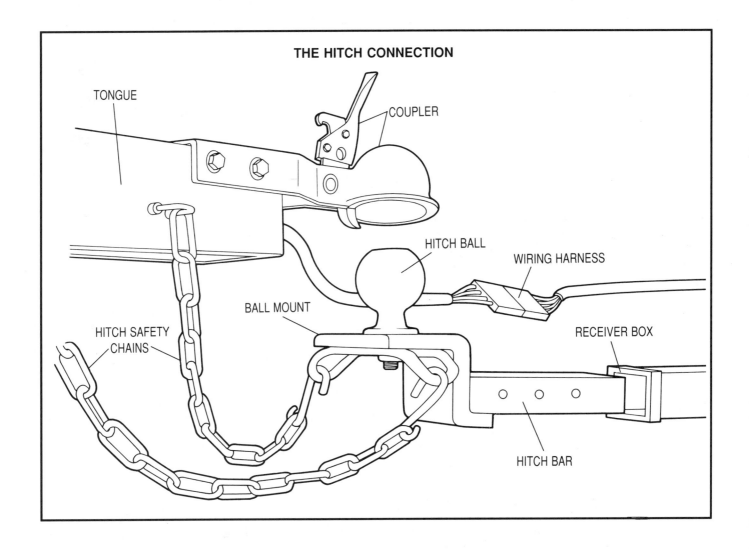

Based on this rating, hitches are broken into four classes:

Class I—This is the light-duty category with a maximum capacity rating of 2,000 pounds. Typical hitch attachment points are to the bumper plus two points on the frame. Common trailer loads in this category are sail and powerboats that are less than twenty feet in length, plus canoes, kayaks, inflatables, and personal watercraft.

The link between the trailer and the tow vehicle is handled by the hitch.

Class II—This is the medium-duty category with a maximum capacity rating of 3,500 pounds. Typical hitch attachment points are to two or more points on the frame. Common trailer loads in this category are sail- and powerboats up to twenty-four feet in length and trailers that hold multiple canoes, kayaks, and personal watercraft.

THE HITCH

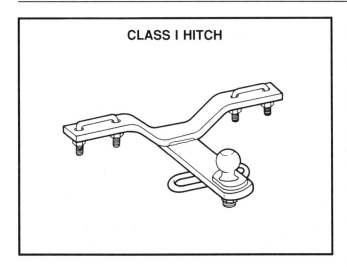

With few frame attachment points the Class I hitch is for light towing up to 2,000 pounds.

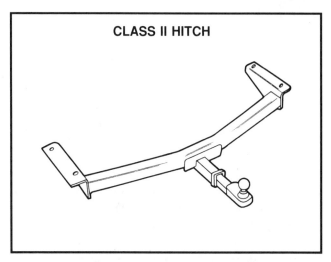

The Class II hitch is for medium towing up to 3,500 pounds. This one is shown with a removable receiver box.

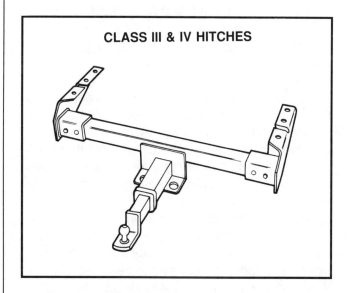

Class III—This is the heavy-duty category with a maximum capacity rating of 5,000 pounds. Typical attachment points are four or more on the frame. Common trailer loads are sail- and powerboats from twenty-four feet to twenty-eight feet in length.

Class IV—This is the extra-heavy-duty category with a maximum capacity rating of 10,000 pounds. Typical hitch attachment points are four or more on the frame. Common trailer loads are large sail- and powerboats.

For heavy towing up to 5,000 pounds, the Class III hitch with its multiple frame attachment points is required. For heavier towing up to 10,000 pounds, a stouter Class IV hitch is required.

HITCH TYPES

In addition to the four weight-rating classes, hitches come in different styles. Light-duty Class I hitches, for example, are offered in four types: 1. a *frame-mount hitch,* which is mounted solely to the frame; 2. a *bumper/frame-mount hitch,* which is mounted to both the frame and the bumper; 3. a *step bumper,* which is a rear bumper with a built-in ball-mount platform or cutaway space to install one; and 4. a *bumper hitch,* which is attached solely to the bumper.

Of these four styles, the frame-mount hitch is the best because it has bolted or welded attachment points on the frame. This helps distribute tongue weight off the bumper and onto the rear axle of the tow vehicle.

The bumper/frame-mount hitch can be useful for light towing, but it has a drawback: It tends to impede the crash-resistant nature of the 5-mph bumper, which is designed to yield slightly upon impact and absorb collision stress.

The step bumper, which is generally found on trucks, is the second-best hitch mount because it, too, relies on solid attachment points on the frame. However, boaters should be aware that there are a number of aftermarket step bumpers that have no frame attachment points. These are useful only for the lightest of towing loads and aren't recommended for trailerboating.

The final light-towing system, the bumper-mount hitch, is the least desirable of the three. Quite simply, bumpers on today's cars aren't designed to handle the loads and stresses of towing. While hitches and trailers are made of steel, modern bumpers and supports are made of a light alloy. Check your owner's manual and you'll find that both automotive and trailer manufacturers strongly advise against using bumper-mount hitches.

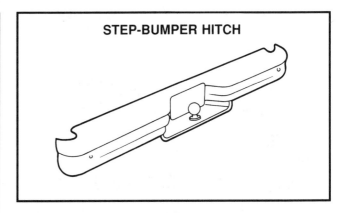

Step-bumper hitches that have attachment points on the frame can carry heavy loads. Those that don't are more for cosmetics than load carrying.

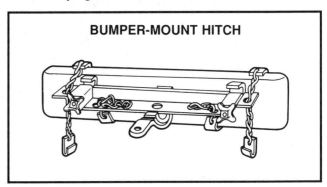

Removable bumper-mount hitches are becoming more rare in towing and aren't recommended for most trailerboaters.

Medium and heavy towing require frame-mounted hitches, and these too come in several styles. Of the variations, two are the most common—the *fixed ball-mount platform,* which is a one-piece unit, and the *receiver hitch,* which has a removable ball-mount platform or hitch bar that slides into a receiver box.

THE HITCH

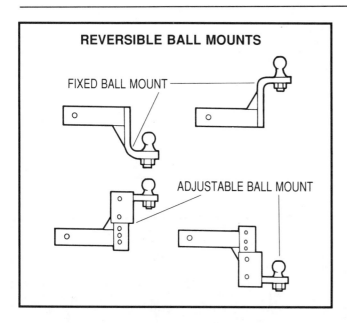

Hitch bars with reversible ball mounts enable the hitch ball to adjust to the height of the coupler.

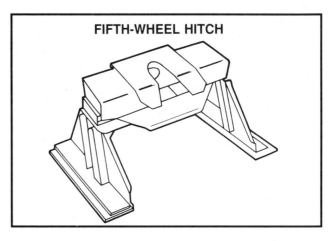

A fifth-wheel hitch is a heavy load–bearing hitch that mounts in the bed of a pickup truck.

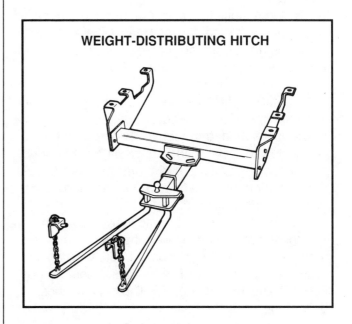

Steel-spring bars on heavy-towing weight-distributing hitches distribute tongue weight to the front of the tow vehicle.

Receiver-type hitches are by far the most versatile of the two. They enable boaters to remove hitch bars and store them in the tow vehicle when not in use. This feature not only forestalls rust but prevents theft. Also, it enables easy adjustment of hitch height. This can be important to boaters with multiple trailers or towing loads. The height adjustment is done through the use of extensions and reversible hitch bars.

Two other variations are fifth-wheel and weight-distributing hitches, which are generally reserved for very heavy loads. Fifth-wheel hitches are mounted in the rear bed of pickup trucks, and except for extremely large and substantial boats, they're rarely used in trailerboating.

Weight-distributing hitches are somewhat more common. These hitches spread out tongue weight through the use of spring bars so that it's shared by all the tow vehicle's wheels. Many automotive manufacturers require the use of weight-distributing hitches, while many trailer manufacturers caution against using them. The best advice is to read your owner's manual carefully before installing one. In general, the automotive industry is a more reliable source of information in this instance.

HITCH BALLS

The component that makes contact with the coupler in a metal-to-metal connection is the *hitch ball*. As its name indicates, this component is basically a sphere that's made of solid steel. There are two-piece hitch balls on the market, but these are inexpensive units and aren't recommended for serious trailerboating.

The hitch ball is fixed to the end of a neck that can be either short or tall depending on the height that's best suited to the trailer. The neck in turn is connected to a mounting shank with a threaded end. A locking washer and a nut are used to secure the mounting shank to a *ball-mount platform* or a *removable hitch bar*.

Hitch balls themselves come in two common diameter sizes: 1⅞ inches for light towing up to two thousand pounds (for Class I hitches) and 2 inches for loads up to five thousand pounds (for Class II and III hitches). For Class IV hitches, a ball diameter of 2⁵⁄₁₆ inches is required. Shanks come in varying diameters, too, generally ⅝ inch, ¾ inch, 1 inch, and 1¼ inch. The wider the diameter, the stronger the shank and the higher the weight rating.

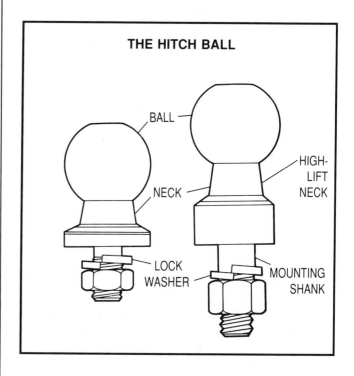

Ball diameter and shank size affect the hitch ball's maximum load rating. Note that a high-lift neck can be used to adjust the height of the hitch ball.

THE HITCH

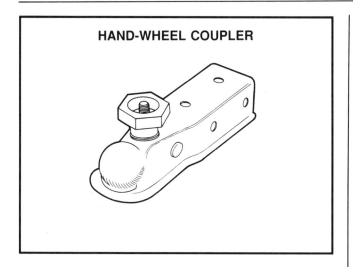

Operation of the ball clamp on this type of coupler is performed by turning the hand wheel. It will accommodate hitch balls of different sizes.

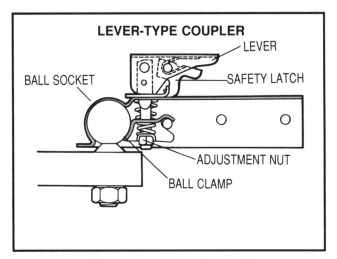

The common lever-type coupler is popular because it has a built-in safety latch. Also, it can be easily secured with a padlock.

THE COUPLER

After the hitch ball, the coupler is the second most important contact point in the metal-to-metal connection between the trailer and the tow vehicle. Located on the very end of the trailer tongue, the *coupler* comes in two basic types—the screw or hand-wheel type and the lever type. Both styles have a coupler socket that fits snugly over the hitch ball and a clamp that locks the ball in place.

The design difference between the two coupler styles is in how the ball clamp is controlled. In one, the ball clamp loosens and tightens by turning a screw knob or a hand wheel. In the other, the clamp is controlled by raising or lowering a lever. Either style can accommodate typical trailerboat loads, but the lever type is more common. Both styles have adjustable locking nuts that alter the tension of the clamp on the ball for fine tuning.

Lever-type couplers provide an extra margin of safety through the use of a trigger-lock mechanism, which probably accounts for their popularity. When the lever handle is pushed down to tighten the clamp, a separate safety device catches to hold the lever in the down position. This trigger lock must be released before the lever can be flipped up again to loosen the clamp.

SAFETY CHAINS

While the coupler incorporates a well-designed locking system, modern hitches also provide a backup mechanism for extra safety. This failsafe role is handled by two *safety chains* that are fitted with hook ends. The safety chains are attached to the trailer tongue or coupler, and are an added insurance that your trailer won't become detached from the tow vehicle while under way. The chain hooks are slipped through rings or holes on either side of the hitch ball.

No trailerboater should ever start down the road without hooking up the safety chains. The minimum breaking strength for these chains should be 2,000 pounds for Class I trailers, 3,500 pounds for Class II, and 5,000 pounds for Class III. For Class IV, it should be equal to the maximum GVWR of the trailer. In general, a 3/16-inch chain is rated to 3,000 pounds, a 1/4-inch chain is rated to 5,000 pounds, a 5/16-inch chain is rated to 7,600 pounds, and a 3/8-inch chain is rated to 10,600 pounds.

Hitch safety chains should be crisscrossed and hooked up for every trip down the road. They should be long enough to prevent binding in turns and short enough to prevent dragging.

THE HITCH

WIRING HARNESS, SURGE-BRAKE CABLE

The final two components in the trailer/tow vehicle link are the wiring harness and the surge-brake cable, which are important not because they bear part of the load, but because they ensure that vital safety features are in working order.

The typical trailerboat *wiring harness* features a flat-four plug that's female except for the ground wire on the tow vehicle and male except for the ground wire on the trailer.

By linking directly to the master brake cylinder, the short *surge-brake cable* or chain is an emergency stop system in case your trailer becomes separated from the tow vehicle while running. The cable pulls a lever/plunger assembly forward and this action triggers the emergency brakes. This failsafe system is hooked by hand to the hitch, and it's something you don't want to forget.

The flat four-prong wiring harness is ubiquitous in boating.

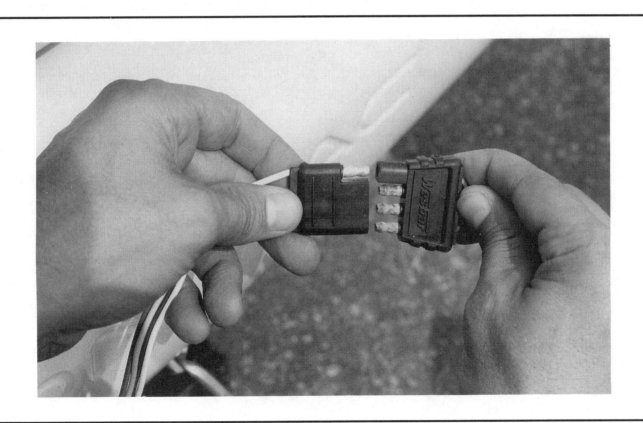

MATCHING THE HITCH TO THE TRAILER

As with selecting the proper trailer, there are many factors to weigh in choosing the right hitch. Here are some tips and guidelines to follow:

- The first consideration is determining the proper hitch class, which is based on the GVWR of your trailer. A Class I hitch has a maximum rating of 2,000 pounds. A Class II hitch has a maximum rating of 3,500 pounds. A Class III hitch has a maximum rating of 5,000 pounds. A Class IV hitch has a maximum rating of 10,000 pounds. Keep in mind that it's not wise to approach the maximum rating too closely. Give yourself a generous cushion, even if it means moving up a class.
- Perhaps the best recommendation for choosing a hitch is to pick one that's strong enough to match the maximum rating for your tow vehicle, even if it's heavier than the GVWR of your trailer. By doing this, if you buy a bigger boat before replacing your tow vehicle, you won't be forced to replace the hitch.
- Carefully consider how the hitch attaches to the tow vehicle. The frame-mount hitch is the best type, because it has attachment points on one of the sturdiest components of your tow vehicle—the frame. A step-bumper hitch is also good, but only if it has frame attachment points. A bumper-frame hitch can work for light towing, but it may negate the impact-absorbing design of your bumper. A bumper-mount hitch is only acceptable for extremely light towing loads and isn't recommended for trailerboating.
- Receiver-type hitches with removable ball-mount platforms offer trailerboaters great versatility compared to fixed ball-mount platforms, but both get the job done. Removable hitches can be stored when not in use, and can be fitted with reversible extensions to facilitate height adjustment.
- Weight-distributing or load-equalizing hitches are often used with very heavy loads. Some automotive manufacturers require them for certain load levels. Be sure to check your owner's manual before installing one.
- Solid-steel hitch balls are recommended for all types of trailerboating. Two-piece balls may be safely used for light loads. For loads up to 2,000 pounds a 1⅞ inch hitch ball can be used. For loads up to 5,000 pounds a 2-inch hitch ball is required. For loads beyond 5,000 pounds, stouter-diameter balls and ball components are required.
- When you set up your towing rig, make sure the ball and coupler are at approximately the same height. This will prevent an excessive angle from occurring that may leave your boat or tow vehicle unbalanced and susceptible to damage. Fine height tuning can be done through the use of reversible receiver-type hitches, ball-mount extensions, and replaceable ball necks.
- Lever-type and hand-wheel-type couplers are equally suitable for trailering, but the lever type has a trigger-lock mechanism that provides an extra margin of safety.
- Finally, be sure to pay close attention to all identification labels and operating manuals for data about load ratings. Virtually every component associated with the hitch and coupler has a load rating. Know what it is and make sure the maximum limit is never exceeded.

CHAPTER 4

THE TOW VEHICLE

It wasn't so long ago that trucks were strictly beasts of burden. Today, their appeal has greatly expanded due to the growing popularity of sport/utilities and minivans. But despite the recent boom in muscular vehicles, modern times haven't been kind to towing. The advent of front-wheel drive, unitized-frame construction, and downsized engines has made a third of all passenger cars no longer suitable for towing. The bulk of the remaining two thirds are rated to tow less than 3,000 pounds (generally much less), and only a handful can tow more than 3,500 pounds.

However, there are some positive developments, too. As noted, many more people are driving sport/utes and minivans than ever before, and sport trucks are also becoming increasingly popular. This boom is actually expanding the towing option to many new people. So while it's not the best of times for tow vehicles, it's not the worst of times either.

What makes a good tow vehicle for a trailerboat? Well, there's no simple answer. A manufacturer's tow rating is based upon analysis of a number of factors, such as vehicle weight, engine size, suspension strength, rear-axle ratio, cooling capacity, transmission, drive system, frame strength, and several others. To better understand a vehicle's tow rating it's important to know about the components and systems that directly affect it.

TOWING CAPACITY

The first thing a trailerboater needs to know about a tow vehicle is its maximum *tow rating* or towing

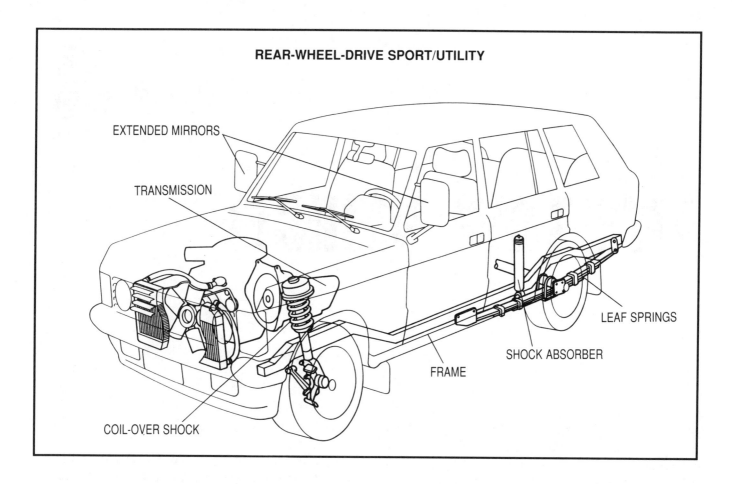

capacity. The place to look for this and other important trailering information is in the owner's manual.

Car and truck manufacturers often list a simple maximum tow rating, but also state a more precise criterion called *Gross Combined Weight Rating (GCWR)*, which includes the weight of both the vehicle and the trailer. This rating takes into account several variables that affect the vehicle's ability to handle a trailered load, such as the weight of passengers, luggage, equipment, gear, and so forth.

To determine the GCWR of your rig, you need to know the weights of the following: the tow vehicle (usually listed as curb weight), the trailer, the

Few modern passenger cars can do more than light towing. The typical tow vehicle today is either a sport/ute, minivan, or pickup, and all are enjoying booming popularity.

boat (including all add-on features, fuel, and water), all passengers, everything carried in the boat (equipment and gear), and everything carried in the tow vehicle (luggage). For some tow vehicles, this exercise in itemizing weight produces some very interesting results.

THE TOW VEHICLE

For example, the first thing you notice is that curb weight has very little to do with a vehicle's tow rating. Some cars and trucks are rated to tow as little as 25 percent of their curb weight while others are rated to tow significantly more. In addition, you'll find that the maximum tow rating of some vehicles allows for surprisingly little additional weight for passengers and gear. In some cases, if the weight of your boat and trailer reaches the maximum tow rating, the GCWR allows for less than six hundred pounds for all passengers and gear. This is approximately the weight of a family of four without luggage.

Most owner's manuals also list a figure for *Gross Axle Weight Rating (GAWR),* which is the maximum load that each axle is rated to carry. Since GAWR figures are generally very close to a fifty-fifty split between the front and rear axles, this measurement doesn't become a major factor for most trailerboaters, except for those with fifth-wheel hitches and rigs with heavy tongue weights.

ENGINE POWER

A popular phrase among gearheads is "You can't beat cubic inches." This refers to the size of the cubic-inch displacement (CID) of the cylinders within the engine block. This component plays a pivotal role in the engine's production of horsepower and *torque,* which is a measurement of how much weight an engine can move—both how far and how fast—by developing a twisting force on the drive shaft.

For trailerboaters, the production of torque is a more important criterion in determining a vehicle's towing capacity than the production of horsepower, especially at low rpm. An engine can be fine-tuned to deliver high horsepower in many ways that don't significantly increase torque. A good example is turbocharging, in which power from the engine's exhaust is used to increase horsepower. Unfortunately, it does very little to improve torque. In general, the role played by an engine's horsepower is higher speed. The role played by torque is more powerful acceleration.

Nothing produces more torque for trailering than an engine block with large CID, which is the ideal kind of engine for a tow vehicle. Tow vehicles need large amounts of torque to accelerate under load, to maintain highway speeds, to climb hills, to achieve passing speed without strain, and to operate efficiently at high altitudes.

The size of a tow vehicle's engine, especially its CID measurement, is a major factor in a manufacturer's overall tow rating. Generally speaking, the bigger the better. However, it's important to note that modern engines, which are typically smaller than their older counterparts, are getting stronger as technology improves. One day, the adage that "nothing beats cubic inches" will no longer be true, but that day is still in the future.

SUSPENSION

As on the trailer, the tow vehicle's suspension system plays a significant role in trailering, especially the rear suspension. Trailer tongue weight is transferred directly onto the rear suspension, and its strength is a major determinant in a manufacturer's tongue-weight and tow ratings.

To beef up the rear suspension, tow vehicles are often equipped with heavy-duty leaf springs, shock absorbers, stabilizer bars, A-arms, helper springs, trailing arms, overload springs, and air bags. With the addition of these rugged compo-

HEARST MARINE BOOKS TRAILERBOAT GUIDE

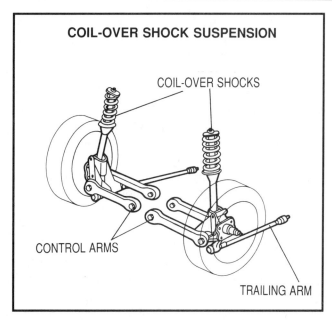

Combining the best of coil springs and shock absorbers, coil-over shocks with control and trailing arms are an alternative to leaf springs for a tow vehicle's rear end.

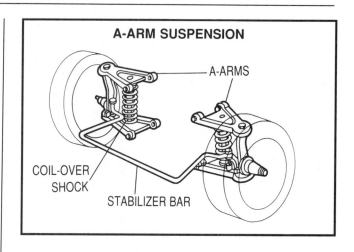

Another rear-end alternative is the A-arm suspension with a stabilizer or antisway bar for added stiffness.

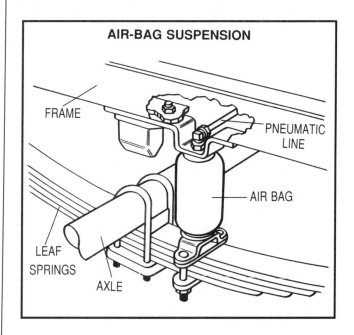

Adjustable air bags act as stiffening components to rear leaf springs.

nents, the tow vehicle's tongue-weight and tow ratings can be increased substantially.

The front suspension doesn't necessarily play a pivotal role in trailering, so adding heavy-duty components here won't have much of an effect on the vehicle's towing capacity. However, it makes sense to have a comparable suspension set up on both ends of the frame, so whenever the rear is beefed up, heavy-duty *coil springs,* shocks, *coil-over shocks,* stabilizer bars, and *torsion bars* are generally added to the front.

THE TOW VEHICLE

AXLE RATIO

Although few nontowers pay attention to *axle ratio*, this is one of the most important variables in trailerboating. The axle ratio is a figure that represents the number of engine revolutions made for each revolution of the drive axle. For example, a mid-range axle rating listed as 3.42 stands for 3.42 engine revolutions (or revolutions of the drive shaft) per revolution of the drive axle. Since most tow vehicles are rear-wheel drive, this specification is often referred to as the rear-axle ratio.

The axle ratio is controlled by a set of ring gears found in the differential. By changing these gears, you can directly influence torque where it's needed most—at the drive axle. With the drive shaft turning more than three times faster than the rate of the drive wheels, the engine builds up a great amount of torque in the drive line. The higher the axle ratio, the faster the engine turns. The faster the engine turns, the more torque it produces. The lower the axle ratio, the less torque.

Manufacturer tow ratings are directly affected by axle ratio. A sport/utility vehicle with a 2.73 axle ratio, for example, may only be rated to tow two thousand pounds. The same sport/ute equipped with the same engine and a 3.42 axle ratio may be rated for four thousand pounds. Put a 3.73 axle ratio on the vehicle and it may be rated for as much as six thousand pounds.

Despite the fact that cars and trucks with higher axle ratios make superior tow vehicles, there's a downside, too. Vehicles with high axle ratios have louder engine noise, lower top-end speed, and poorer fuel economy. For these reasons it's a good idea to make sure your choice of axle ratio is the right one for your trailering needs.

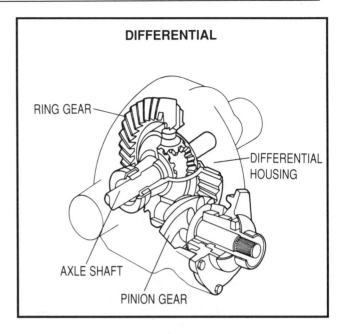

Inside the differential housing, the torque of the drive shaft is transferred to the rear axles through the pinion and ring gears.

DRIVE SYSTEMS

It makes a big difference to trailering and tow ratings if your vehicle is front-wheel drive (FWD) or rear-wheel drive (RWD). The same is true for two-wheel drive (2WD) and four-wheel drive (4WD). Here's why.

Instead of the typical drive shaft that's used in rear-wheel drive, a front-wheel drive system uses a transaxle. A *transaxle* is an assembly that combines the transmission, differential, and drive axle into one unit. Because transaxles are designed for

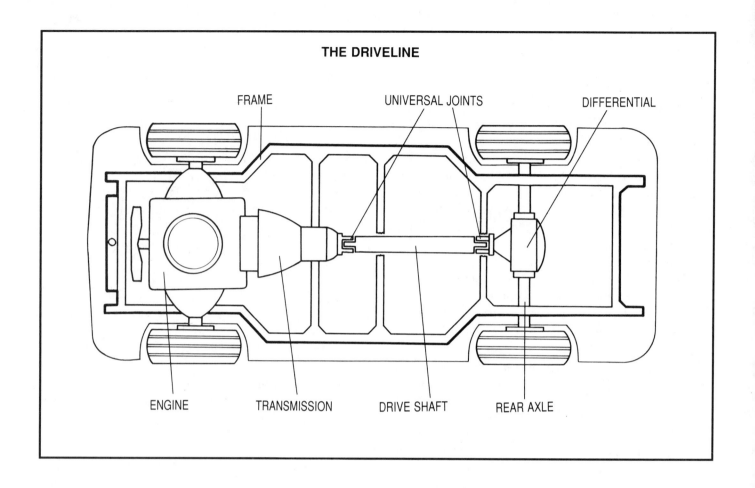

use in passenger cars, they're only built for light-duty use and generally don't possess the same strength as RWD systems. Consequently, FWD systems have lower tow ratings than RWD systems, typically much lower.

There's another drawback to FWD for towing, and you encounter it on the launch ramp in the form of reduced traction. The problem is caused by tongue weight pressing down on the rear end and lifting up the front. The effect is compounded during acceleration, which also tends to shift vehicle weight to the rear. The result is a sharply reduced capacity to pull a boat out of the water, especially up steep, slippery ramps.

The components that transfer power from the engine's crankshaft to the drive wheels comprise the driveline. Shown here is a conventional rear-wheel-drive vehicle.

Another potential problem for FWD occurs while under way. If the tongue weight raises the drive wheels, even slightly, the result could be reduced steering control and trailer sway. For these reasons, FWD vehicles are generally rated for light-duty towing only.

THE TOW VEHICLE

The benefit of four-wheel drive is improved traction in all conditions, which is an important feature to have on wet launch ramps. Once the 4WD system is engaged, power and torque are provided to all four wheels. This is accomplished by a series of gears, which in addition to delivering power also provide a high and low range of operation. When the low-range gears are engaged, engine rpm increases. The result is high torque production, which comes in handy when negotiating difficult conditions.

Despite the obvious benefits of 4WD, it isn't the answer for every boater's needs. In some cases, manufacturers lower tow ratings for 4WD vehicles. The systems add weight to the vehicle, reduce top-end speed, and potentially increase maintenance costs.

Two other drive systems that can affect trailering are all-wheel drive (AWD) and limited-slip differential. All-wheel drive is similar to 4WD except that it's engaged full-time. The benefits and drawbacks of AWD are similar to those of 4WD.

Limited-slip differential is designed to prevent one drive wheel from spinning uselessly while the other remains motionless. It does this by partially locking the right and left axle assemblies so that both turn and deliver power. Limited-slip differential is available on both 2WD and 4WD vehicles. Its benefits and drawbacks are similar to those of 4WD.

TRANSMISSIONS

Quite often experienced drivers are puzzled when they learn that manufacturers' tow ratings are generally downgraded for manual transmissions compared to automatics. There are several reasons for this. The first is that a manual transmission requires the driver to engage a clutch and physically shift between gears. Although it gives the driver tremendous control over the drive train in most situations, during trailering it puts more stress on the components than they were designed to absorb. Manufacturers don't believe that clutches in passenger cars and light-duty trucks are durable enough to handle towed loads, especially when negotiating steep launch ramps.

In addition to an automatic transmission's ability to shift gears automatically, it has a torque-sensing capacity that makes it ideal for adjusting to the added strain of a trailerboat. Also, drivers with automatic transmissions don't have to ride the clutch to build up torque when it's needed for pulling power.

COOLING SYSTEMS

Nothing is more potentially damaging to hard-working vehicle parts than heat, and nothing builds it up faster than pushing an engine to its limit by towing a heavy boat. To control heat buildup, typical passenger vehicles are equipped with light-duty cooling systems, which aren't suitable for towing.

A good tow vehicle's cooling system will feature beefier parts. These include a radiator with additional core layers, a radiator fan with additional blades and a thermostatic clutch, a high-performance water pump, an auxiliary transmission cooler, and an engine-oil cooler. The addition of these and other cooling-system parts will greatly improve a vehicle's tow rating.

MANUFACTURER'S TOW PACKAGE

Perhaps the biggest single factor affecting a vehicle's tow rating is whether or not it has a *manufacturer's tow package.* This cluster of factory-installed trailering components is an option package available at the time of purchase and can more than double the vehicle's overall tow rating.

Here's a rundown of what's included in a typical manufacturer's tow package: heavy-duty radiator and fan, transmission or transaxle cooler, engine-oil cooler, high-performance water pump, heavy-duty turn signals, heavy-duty suspension system, factory-installed hitch, higher-ratio axle gearing, heavy-duty front brakes, high-amp alternator, heavy-duty battery, factory-installed wiring harness, and extra-wide exterior mirrors.

A manufacturer's tow package, such as the one listed above, will generally cost several hundred dollars when the vehicle is ordered. To make the changes at a later time will probably cost several thousand dollars.

CHOOSING A TOW VEHICLE

Factors like comfort, style, performance, handling, and price are important considerations when buying a new vehicle. When the vehicle is intended for trailering, a whole new set of factors must be added. Here are some tips to keep in mind when making your evaluation:

- The first three things you need to know about a tow vehicle are its tow rating, tongue-weight rating, and GCWR. Since you already know the weight of your boat/motor/trailer, set this figure as your vehicle's minimum tow rating. If you haven't purchased your trailering rig yet, make your vehicle's maximum tow rating the upper limit for your rig. Be sure to leave a cushion of at least several hundred pounds in your vehicle's tow rating and GCWR for safety. Remember that tongue weight should be no more than 10 percent of your rig's total weight, and that the weight of passengers and luggage must be added into GCWR.
- When you're presented with an engine option, the bigger the better. While under load, big engines strain less than small engines, and engines that labor least last longer. Also, remember that production of torque is more important than production of horsepower.
- Car or truck? Few passenger cars today can handle more than light towing, and only a handful of cars can tow more than five thousand pounds.
- The thing to note about a rear suspension is how it reacts to tongue weight. Make sure it provides a level ride without lifting the front end and is sturdy enough to prevent bottoming out over bumps.
- While higher axle ratios are generally beneficial to tow vehicles, they're not the answer in all cases. A balance must be struck between the axle ratio, the size of the engine, and the weight of the towed load. Fortunately, manufacturers make this choice easy by charting tow ratings based on different axle ratios and engine sizes. One tip to keep in mind is that little will be gained by choosing an axle ratio so high that it gives you twice the towing capacity you need. In fact, a great deal may be lost in terms of nontowing performance.
- Front-wheel drive or rear-wheel drive? Basically, FWD systems are built and designed for use in passenger cars. Consequently, they're not as heavy-duty as RWD. Also, the front wheels in FWD are subject to poor traction

THE TOW VEHICLE

when weight is shifted to the rear during the normal course of trailering. However, FWD is suitable for many light and medium towing situations.
- Two-wheel or four-wheel drive? Some manufacturers lower tow ratings slightly for vehicles equipped with 4WD; questions about durability under towing stress for long periods of time are probably the reason. Another important consideration is that 4WD adds weight to the vehicle and generally requires higher maintenance costs. However, nothing beats 4WD in northern climates and on steep, slippery launch ramps.
- Manual or automatic transmission? The clutch is the weak link in a manual transmission. Most manufacturers don't believe it can handle the strain of trailerboating. For this reason, tow ratings for manual transmissions are often significantly lower than for automatics.
- There's very little controversy surrounding manufacturer tow packages. Order one at the time you purchase your vehicle and it will cost several hundred dollars. Order it later and it will cost several thousand dollars.

CHAPTER 5

ON THE ROAD

Trailering is a white-knuckle affair to many people, but it doesn't have to be. In fact, if both your trailer and tow vehicle are properly set up, you'll hardly know you're pulling a load. But what's a proper setup? Much of this information has been covered in previous chapters, but in this section it will be pulled together in the form of tips for pretrip safety checks and on-the-road driving techniques.

You've probably heard war stories or seen firsthand evidence of trailering trips gone awry. In a worst-case scenario a trailer can snake all over the road and drag you places you never intended to go. The boat can shift on the cradle and become dangerously unbalanced. The surge brakes can malfunction and force you into a jackknife. And worst of all, the trailer can become unhitched and take off on its own.

As with most endeavors in life, preparation is the key to success. Simply stated, know before you tow. The best way to do this is to practice backing up, braking, trailer control, and overall maneuvering in advance of your trip. Then check your boat, trailer, and tow vehicle before heading out. Do this before each trip, no matter how short. Between trips, follow a regular maintenance routine. And then, when you're on the road, drive with an attentive eye and ear for possible trouble. If you do this, your trailerboat trips will be memorable for all the right reasons.

TOW VEHICLE CHECKUP

Every vehicle requires a certain amount of care and feeding, but none more so than a tow vehicle.

This is especially true if it's pulling close to its maximum tow rating. Before each trip open the hood and check all fluid levels, including oil, coolant, transmission fluid, power steering, windshield washer, and battery (if necessary). Then check all hoses for pliability, cracks, and leaks. Check all belts for wear and proper tension. Check the air filter for dirt.

Moving around the vehicle, check the tires for tread wear and proper inflation pressure. And don't forget to do the same for the spare. One by one, turn on the headlights, signal flashers, and brake lights. Make sure all are working. Check the fuel level and fill up the tank before connecting the trailer. However, if your boat needs fuel too, it's a good idea to fill both at a local gas station. Gas prices are generally lower on the road than on the water. Finally, make sure the hitch and hitch ball are wrench tight.

After hitching up, check the road clearance beneath the back bumper. Make sure it's high enough to avoid bottoming out over bumps. Finally, make sure the tow vehicle and trailer are level or close to it when connected.

TRAILER CHECKUP

Start at the hitch and make sure the coupler latch is properly seated against the ball. Make sure the locking lever is in the down position, and the safety pin is in place. Also, if it's a receiver-type hitch, make sure the locking bolt is in place and that it's secured by a cotter pin. If the equipment is new, make sure the ball is the right size for the coupler, and make sure all components are rated to tow the load. Finally, check all nuts and screws on the coupler to make sure they're wrench tight.

Check the trailer's electrical connection and make sure it's tightly plugged together. For long journeys, it's a good idea to wrap electrical tape

THE EFFECT OF TONGUE WEIGHT

Driving with too much tongue weight will cause mushy steering and possible damage to the tow vehicle. Driving with too little tongue weight will result in serious trailer sway. In addition to having the proper tongue weight, the trailer and tow vehicle should line up for a level ride.

ON THE ROAD

around the plug. This ensures that the plug stays connected and guards against rain. Also, make sure all the wires are in good condition and that they're secured against chafing or dragging en route. Every other trip, make sure all nuts and screws on the trailer are wrench tight.

At this point, go to the tow vehicle and test the entire trailer lighting system. Before moving on, make sure that the safety chains are crossed and securely hooked to the hitch. Also, make sure there's enough slack to prevent them from binding in turns, and conversely, not enough slack to allow them to drag on the ground. Finally, make sure the breakaway lanyard is attached to the hitch, and that it, too, has the proper amount of slack.

Make sure the tongue jack has been fully raised, and if the dolly wheel or stabilizer jack is removable, make sure it's stowed for the journey. Check the winch and make sure the strap hook is tightly secured to the boat's bow eye and the brake lever is locked. Also, make sure the safety chain is hooked to the bow eye and the boat is snug up against the bow stop.

Then go to the tires and wheels. Check the tires for tread wear and proper inflation pressure, and don't forget the spare. Check the wheels to make sure the lug nuts are tight and that none are missing. Finally, check the wheel bearings to make sure all are filled with grease.

Before moving on to the boat itself, check the tie-downs to make sure they're snug and well hooked to the trailer. Take special note of any contact points that might result in chafing, and insert pads where necessary.

BOAT CHECKUP

Walk around the trailer and check how the boat rests in the cradle. Make sure it's level from side to side and from front to back. Look under the hull to see that the strakes and chines are properly positioned on the bunks or rollers. Also, be sure that bunks or rollers are directly beneath the transom for support. Remember that the transom should neither extend beyond the bunks or rollers nor sit far forward of them. Except in the case of lightweight loads, such as personal watercraft, canoes, or small sailing skiffs, the trailer should be a virtual perfect fit. Lower antennas, vertical-mounted fishing rods, and other tall objects to avoid surprise encounters with overhead obstructions. Make sure the engine's drive leg is tilted up and out of the way.

Climb inside the boat and make sure that nothing can come loose. This includes the rigging, gear, hatches, and especially the battery. Make sure that the stowed gear keeps the center of gravity toward the bow. Is all Coast Guard safety equipment aboard? Do you have all the necessary keys?

Check the fuel and oil levels and then decide whether you should replenish on the road or at the marina. Don't forget to tighten all lids. After this, you're ready to put on the canvas cover and securely tie it down. It's always a good idea to drive with your boat completely covered up. This not only protects the boat, but economizes fuel. Also, check the transom tie-downs, which are good accessories to have for every trip.

BASIC DRIVING TECHNIQUES

Trailering isn't hard to do, but it requires more work than simply driving a car. So it's a good idea to drive only when you're rested, and on long trips, plan to log fewer miles per day than you normally would. It also requires a serious adjustment of basic driving skills. Any time a trailer is hitched to a

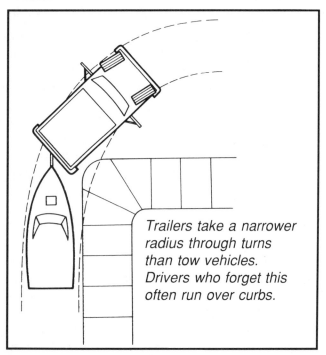

Trailers take a narrower radius through turns than tow vehicles. Drivers who forget this often run over curbs.

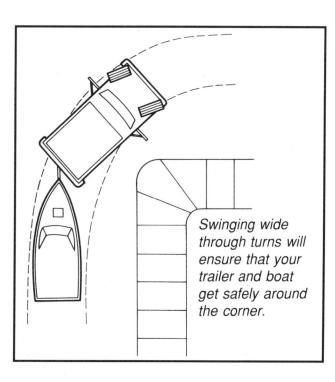

Swinging wide through turns will ensure that your trailer and boat get safely around the corner.

vehicle normal handling characteristics are altered, and the driver must respond accordingly. The most important tip is to allow yourself more time to react, and the best way to do this is to slow down.

Slower speeds give you more time to react to changing road conditions. With several thousand pounds attached to your backside, you need every second you can get. Allow at least one length of your complete rig (tow vehicle plus trailer) for every 10 mph. Stay in the right and middle lanes. Note any sluggishness in acceleration, and allow extra space for merging and passing.

Generally speaking, develop a good overall sense of anticipation. Begin reacting to posted hazards, such as sharp curves, merging traffic, narrow bridges, detours, and bumps as far in advance as possible. Now more than ever it's important to observe conditions as far down the road in front of you as possible.

A tip to keep in mind that's related to speed has to do with the overdrive setting of an automatic transmission. The tip is to stay out of it, unless you can maintain a steady speed on a limited-access highway. The engine generates relatively little torque in overdrive and manufacturers recommend against using it.

Aside from slowing down, moving safely through sharp corners is the next most important task. Remember that you must swing wide through turns to accommodate the extra length of the trailer. Don't begin turning the steering wheel until you're slightly beyond the corner. If you forget, the trailer may run over a curb or worse.

Braking is the next most important consideration. Even after leaving extra space between you and other traffic, keep in mind that towing a trailer dramatically increases the distance required to stop. This is true despite the fact that your trailer may be equipped with surge or electric brakes.

Ideally, when you need to use the brakes you'll

ON THE ROAD

have time to apply them properly. Tap the brake pedal gently and then pump it harder as you gradually slow down. This will ensure a smooth, predictable stop. Even in emergency panic-stop situations, your trailer is designed to maintain a safe, predictable line behind your vehicle.

Still, anything can happen in an emergency, including the unexpected occurrence of a poorly secured boat climbing forward over the top of a winch stand. So give yourself the best chance for success by making sure the trailer brakes are properly adjusted. They should be adjusted neither too strong nor too light. Too strong means that they lock up whenever steady pressure is applied to the brake pedal. Too light means that they don't kick in until you slam the brake pedal to the floor.

Trailer brakes should be activated between these extremes at the point where hard braking begins. Interestingly, when well-adjusted trailer brakes are activated, they actually help slow the tow vehicle from the back end, and give the driver an enhanced feeling of control. Poorly adjusted trailer brakes, on the other hand, can result in jackknifing.

And finally, take special care in climbing and descending hills. Hill climbing causes your engine, transmission, differential, and wheels to generate more than normal heat. Watch your gauges carefully up long hills, especially during hot weather. On the downside of the hill, shift into lower gears instead of using your brakes to control descent. Remember that overheated brakes are subject to fade.

BACKING UP

The first piece of advice to keep in mind when backing up with a trailer is to relax. The second is to practice, especially with a friend who can act as an observer.

To some trailerboaters, backing down a launch ramp is the single most stressful time of the trip. It doesn't have to be. Start by building the necessary skills the right way. During off hours, go to an empty parking lot and spend some time getting the feel of your rig. You can set up test courses by using existing painted lines or by bringing plastic traffic cones.

At first, simply get a feel for the rig's turning radius, its length, width, and the new experience of relying on the side mirrors, especially the tricky right side mirror. Start off by making several turns in an increasingly tighter radius. Learn how the trailer follows the tow vehicle. Then, carefully and slowly, turn the steering wheel fully to the left and right. Find out the points where the angle between the tow vehicle and trailer are so sharp that they make contact with each other. Mentally note these points and avoid them to prevent vehicle and trailer damage.

Now it's time to assume the basic backing-up hand position, which means that you should place one hand and one hand only on the bottom of the steering wheel. With your hand in this position, move it in the direction you want the trailer to go. Forget about the tow vehicle for a second, and just think about the trailer. Do you want the trailer to go right? Then turn your hand to the right. Do you want the trailer to go left? Then turn your hand to the left. It's that simple, at least in theory.

A special note to owners of trailers equipped with surge brakes: Don't keep your foot pressed on the throttle in reverse if the engine's racing and you're not making any progress. Poorly adjusted surge brakes are sometimes activated when the trailer is backed up, especially when going uphill. Modern surge brakes are designed to prevent this problem from occurring, but an occasional adjustment may be needed.

HEARST MARINE BOOKS TRAILERBOAT GUIDE

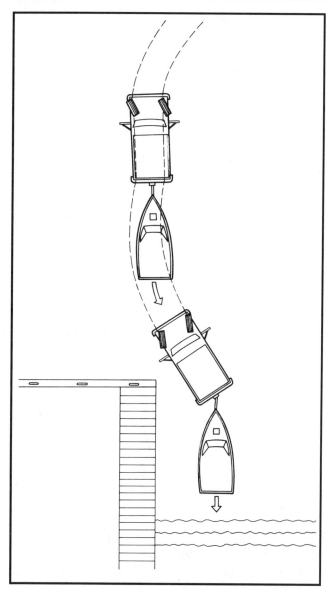

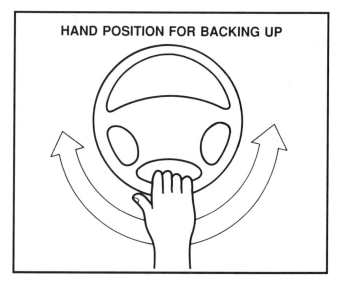

To make it as easy as possible to learn to back up with a trailer, start with your hand resting on the bottom of the steering wheel. If the trailer needs to go to the right, rotate your hand to the right. If it needs to go to the left, rotate your hand to the left.

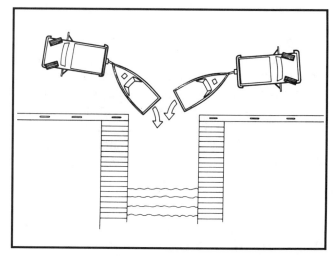

It's virtually impossible to back up any distance without making midcourse corrections. In fact, backing up is often best described as a series of small S-turns.

One good backing-up tip is to maneuver the rig so that you back in from the left. Why? You get a full view of the situation out the driver-side window.

ON THE ROAD

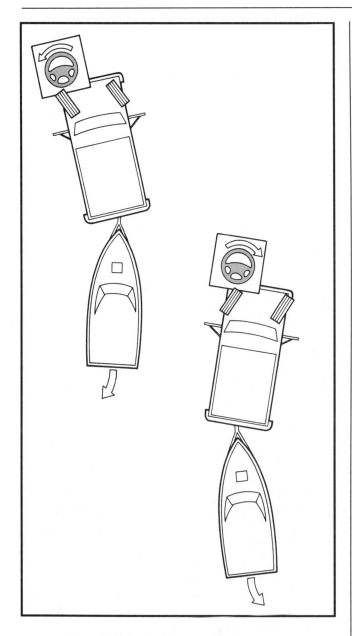

The effect of turning the steering wheel and maneuvering the front wheels while backing up a trailer runs counter to typical driving experience. Take the time to practice in an empty lot before heading to the launch ramp.

On the launch ramp, numerous factors come into play and a couple of hours of practice will help you cope with them. Set up several courses that enable you to get familiar with common launch-ramp scenarios, such as backing up in a long straight line, backing up around a lefthand and a righthand corner, and backing into tight spaces.

With your hand placed confidently at the bottom of the steering wheel, the first thing to remember is to go slowly. This will allow you to minimize wandering and make timely corrections. Push the throttle gently and move your hand in the direction you want to go. Once under way, you can move your hands to a comfortable position and make small adjustments to enable the tow vehicle to follow the trailer. However, if the trailer starts heading where you don't want it to go, then immediately stop and begin again.

One good tip to remember is to set yourself up whenever possible to back the trailer to the left. Why? The reason is that the driver's seat is located on the left and you can get a full view of what's happening simply by turning around and looking out the window. When backing to the right, you're forced to rely on the right side mirror, which requires practice. You won't be able to back up to the left every time, but do it when you can. It tends to make life easier.

Finally, take note of how much input the steering wheel requires to turn the trailer. In general, short trailers require less steering input than long ones. The deciding factor is the distance between the trailer axle and the hitch. Short trailers tend to jackknife easily. Long ones tend to be more forgiving. The best advice is to take an easy-does-it approach and provide no more input than necessary. If a series of hard corrections is required, your best advice is to start over again, because at that point, you're probably on the wrong course anyway.

MEETING SPECIAL CHALLENGES

Drivers are always urged to be aware of everything that goes on around them, to anticipate trouble and drive defensively. This advice is especially important for trailerboaters. Learn to keep your head on a swivel. Scan traffic and road conditions in all directions. Be alert. Anticipate. React. As always, prevention is the best medicine.

One area of special concern for trailerboaters is the right side of the vehicle, which can often be a blind spot. To help cope with this problem, make sure the tow vehicle's mirrors are the right ones for the job. They should extend out and away from the tow vehicle far enough to provide an unobstructed view of the boat, which is very important for monitoring the load's progress, plus observing the inside lane of traffic, which is important for switching lanes. This may require the use of supplemental equipment, such as oversize mirrors, extension brackets, and a convex spot mirror. Go this extra mile, even if it's not mandated by law in your state. It could make a big difference in a tight situation.

Among the first things trailerboaters will notice on the road is that automobile drivers seem oblivious to your special needs. In addition to riding in the blind spot, they'll cut you off, tailgate too closely, slow down in the middle of a steep hill climb, and perform a number of other equally challenging maneuvers. Despite this apparent lack of courtesy, trailerboaters need to keep a level head at all times. With a boat in tow your options are limited. Simply adjust to each situation as it arises, and keep making progress toward your destination. Remember, it will all seem worthwhile when you reach the marina.

TRAILER SWAY

To a certain extent, all trailers *sway*, which means they have a tendency to wander from side to side while under way. However, occasional sway and excessive sway are two completely different things. Occasional sway is normal. Excessive sway is the trailerboater's worst nightmare.

A typical example of sway occurs when the trailer is disturbed from its normal course by a gust of wind, a large truck passing, or bumps in the road. During this kind of sway the trailer may lurch and then resume its normal course.

A more troublesome kind of sway occurs when the trailer develops a fishtail motion that remains constant at highway speeds. This is a warning sign that all is not well with your towing rig. It may not worsen to the point of becoming dangerous, but there's no guarantee it won't. If steady sway occurs while you're driving, start planning an immediate course of action. Slow down to a speed that diminishes some of the motion. Do this by letting up on the throttle instead of using the brakes. Then look for a suitable place to pull over and inspect your rig.

The problem with steady sway is that it may eventually worsen into a more serious condition called growing or excessive sway, where the side-to-side motion increases with each swing of the pendulum. In this scenario, the trailer can take control of the tow vehicle with disastrous results.

Both steady and excessive sway are typically caused by tongue weight that's either too heavy or too light, or a load that's too heavy for the tow vehicle. Other causes are a short wheelbase or a long overhang on the tow vehicle, or a flexing frame or soft suspension on the tow vehicle. But before reaching these dire conclusions, it's a good idea to start by inspecting easy-to-fix problems.

Look at the trailer tires first to make sure they're

ON THE ROAD

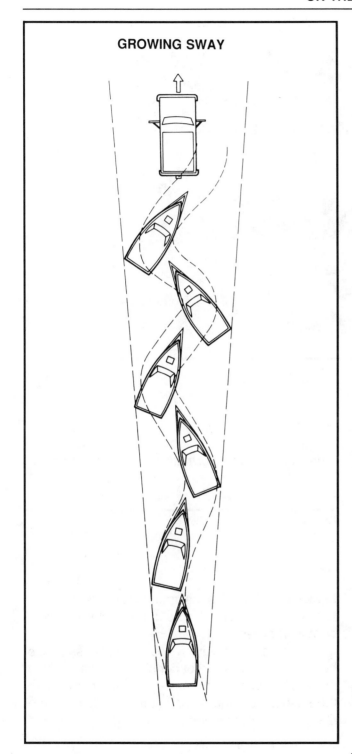

GROWING SWAY

properly inflated. Then inspect the trailer/tow vehicle contact points. Make sure the ball is fully seated and locked in the coupler. Then make sure the coupler is solidly bolted to the trailer tongue. On the opposite end, be certain that the hitch is firmly welded to the tow vehicle frame. For good measure, grab the trailer tongue and hitch firmly, and shake. What you learn by doing this will be important, but it won't necessarily solve the problem.

Move to the trailer now, and check your load distribution. Add-on gear, supplies, equipment, and so forth may have broken loose from their moorings and in turn altered the trailer's balance point. Or they may have been stowed poorly to begin with. Repacking will help, but again, it may not solve the problem.

As mentioned earlier, significant trailer sway is usually caused by two fundamental problems: light tongue weight or a load that's too heavy for the vehicle. These problems are difficult to fix while under way. Check to see that the boat is resting properly on the trailer bed. If it has moved backward while under way, use the winch to pull it back in place. Then check the boat's fuel and water levels. Are the tanks located behind the trailer's axle? Are they more full this time than usual? If so, jettisoning the water may help. Jettisoning fuel must be done by filling dedicated fuel cans.

The serious problem of growing sway occurs when the duration of sway increases with each swing of the trailer. Eventually, a trailer prone to growing sway may take control of a tow vehicle.

Other factors to consider cannot be corrected beside the highway. Is the load unbalanced because of incorrect placement of the trailer's axle? If so, the axle must be moved until it achieves the proper tongue weight. Is the load too heavy for the tow vehicle? If so, beefing up your suspension may help, but only if the problem is excessive tongue weight. You can check your tongue weight by resting the coupler end of your fully loaded trailer on a bathroom scale. Remember, it should be no more than 10 percent of your total load. For the best driving results, a few points less than 10 percent is often recommended. However, keep in mind that a load with insufficient tongue weight will sway, too.

At this point, it's wise to double-check all of your weight ratings and make sure no limits are exceeded. Legally, you cannot tow a load that's too heavy for a vehicle's maximum rating. Is this your fundamental problem? If so, a new vehicle or new boat is in order.

In addition to sway, a load that presses too heavily on the rear suspension will cause two other problems for drivers—potential damage to the suspension and mushy steering, which is the result of the front wheels being slightly lifted into

Maximum weight ratings for the trailer, axle, tongue, coupler and hitch should be double checked against information printed on trailer labels.

ON THE ROAD

the air. Solutions here are stiffening the rear suspension or adjusting the trailer axle to achieve a lighter tongue weight.

One final on-the-road tip. After you've gone through all the pretrip safety checks and you've begun an extended trip, it's a good idea to pull over occasionally to check the rig even if nothing seems apparently wrong. In fact, it's a good idea to do a quick status check of the rig every time you stop. Be sure to check the air pressure and heat of the tires and trailer wheel hubs. If the wheel hubs are excessively hot, it could be a sign of improper lubrication or a signal that the surge brake is engaging frequently. Be sure to inspect the hitch and coupler, too. The sooner you discover a problem, the easier it is to fix.

The problems discussed in this section are not necessarily the norm for trailerboating, but it's a good idea to be aware of them and their potential solutions.

CHAPTER 6

ON THE WATER

If you want a good laugh while learning some important trailerboating tips, just sit for a couple of hours beside a busy launch ramp on a hot summer day, especially on a weekend. What you'll see is a series of slapstick sketches involving boats launched without drain plugs, trailers backed into obstacles, propellers dragged over concrete, and on the rare occasion, driverless tow vehicles plunged into the water.

Is this funny? It depends on your point of view. If your expensive equipment is involved, then it's probably not so rib tickling. But if you're just an observer, then it can be entertaining, because the ultimate joke is that the dents, dings, and drenchings are entirely preventable.

There are two keys to successful boat launching and retrieving: practice and preparation. Practice, as indicated in the last chapter, means going to an empty parking lot and getting the feel of driving with a trailer. Specifically, familiarize yourself with how the load affects engine responsiveness and braking, how it requires adjustment for extra length and width, and how it alters the fundamentals of cornering, backing up, and negotiating tight spaces. After a few hours you'll be ready for step number two—preparation.

PRELAUNCH PREPARATION

The first thing to do after arriving at the launch ramp is to find a place away from the bustle of the main launching area. There's nothing to be gained by heading right for the water; in fact, much could be lost.

Start by taking the tie-downs and canvas cover off the boat and stowing them away. Did you fill the boat's fuel tank before arriving? If not, make your first priority after launching a fuel stop. Did you bring all the necessary gear and supplies? If not, buy and stow them now before heading out on the water. This is also the time to move everything from the tow vehicle to the boat, such as coolers, extra clothing, towels, sunscreen, cassette tapes, and so forth. Performing these tasks now will maximize your time out on the water and minimize it at the launch ramp. It can get hectic on the ramp with anxious boaters waiting in line to use the facilities behind you.

Next, unhook the winch-stand safety chain from the bow eye. However, leave the winch line or strap engaged until the boat's in the water. Insert and tighten all drain plugs. Disconnect the tow vehicle/trailer wiring harness to prevent potential shorting. Then pull two docking lines out and tie them to deck cleats. Finally, check your trailer's wheel bearings to make sure they're fully packed with grease and cooled down, which usually takes about fifteen minutes after the journey.

One of the first pre-launch duties is to replace the transom drain plug. Establish a pre-launch checklist to make your routine on the ramp as swift and easy as possible.

RAMP INSPECTION

You're not quite ready to head for the water, but you're close. Even if you're familiar with the launch ramp, it's a good idea to walk over and give it a visual inspection. Conditions and water levels can change dramatically between boat trips.

Check to see if you can back in a straight line to the water or if you have to come in at an angle. If at an angle, your best approach is to keep the water to your left when pulling in past the ramp. This enables you to back up to the left, and allows you to inspect your progress by looking out the driver-side window. Backing up to the right forces you to rely on the right-side mirror. Before walking down the ramp, find out where you can park the tow vehicle and trailer.

Now use a critical eye to evaluate the ramp itself. Is it steep? Is it slippery? Is it wide enough to accommodate more than one trailer? If so, is one side better than the other? Why? Is there a drop-off at the end of the ramp? Are there rocks or other obstacles in the water? Is there strong current or gusting wind? Is the ramp in a tidal area? If so, how will conditions change throughout the day? Finally, check out the location and conditions of the dock where you'll tie up after launching.

ON THE WATER

LAUNCH PROCEDURE

With tow vehicle and trailer ready, and a boat driver at the helm, the first thing to do is relax. Then assume the correct backing-up hand position on the steering wheel and slowly back down the ramp. Have an observer watch your progress and signal when the trailer and boat are immersed in the water. The trailer is correctly immersed when the boat floats slightly on the aft end. Note the point at which the water level intersects with the trailer wheels and then use this point as a marker for future launchings. For some trailers, the wheels will be halfway submerged. For others, the wheels may be fully submerged.

Now, turn the tow vehicle engine off and use the emergency brake and transmission to hold the vehicle in place. Walk back to the trailer, flip the winch lock lever to the open position, and wind out some line. When the line is slack, unhook the winch line. At this point, especially for those with light craft or roller trailers, the boat can slide off the trailer and tie up at the dock.

If the boat doesn't slide off the trailer bed at this point, don't worry about it. Most modern trailers are designed for drive-on/drive-off use, which means that marine power is required for the final push. When the trailer is in place and the winch line is unhooked, signal the boat driver to press the trim button, submerge the prop, and start the engine. Then with just a bump or two on the throttle, slowly back the boat off the trailer. If the trailer's in top condition and well matched to the boat, the tow vehicle driver shouldn't need to push the boat off. However, a light shove or two doesn't hurt when necessary. Now pull the trailer out of the ramp area.

Two things to keep in mind when using marine power when launching: 1. Be sure to set the trim so that the engine's lower unit is deep enough to

LAUNCHING

When the stern end of the boat begins to float, you can simply drive it off the trailer.

Idle slowly back and trim the motor down as you get into deeper water.

submerge the prop and water pickups, but not so deep that the prop is in danger of striking the bottom; and 2. Make sure the engine is warmed up and running properly before backing off the trailer. You don't want to go dead in the water the second you're out of reach. To do it properly, let the engine run a minute while checking the oil pressure and flow of water through the cooling system.

One special reminder for inboard and sterndrive powerboaters: Before turning the ignition key, don't forget to purge the engine compartment of fumes. In fact, it's a good idea to open the engine hatch and turn on the blower before dropping the boat into the water. This is especially important after long trailer journeys, which may cause fuel to slosh around and build up fumes.

When the tow vehicle and boat go their separate ways, the tow vehicle driver heads for a good parking spot while the boat driver heads for a nearby dock to tie up. Both drivers should move at sensible speeds, but dawdling is discouraged. Remember, others are waiting to use the facilities.

Sound difficult? It isn't. With a little practice the entire procedure—from prepping to launching—takes less than ten minutes.

Launching a sailboat with a deep keel typically requires a crane or hoist, as shown above. Note that the slings are carefully placed on the bow and stern.

LAUNCHING BY SLING

Not all marinas are equipped with launch ramps, especially those located on saltwater. Many of these use a lift or a sling to drop boats into the water. For owners of deep-keel sailboats, this is basically the only way they can be launched.

In general, you should follow the same fundamental advice for using a sling as given earlier for launching on a ramp. Notably, this includes prelaunch preparation and launch-area inspection. However, there are several important differences.

For one thing, sling launching is quite a bit easier on the boat owner than ramp launching. The sling or lift operator directs you to the proper location, and then handles much of the task himself. He runs the slings around the boat, operates the lift, and then unstraps it. However, it's a good idea for the boat owner to lend a hand, especially if you know where the boat's internal support points are located. This can be important information if the boat is heavy and its sides are subject to flex. Another point to keep in mind is that special care should be taken with boats that have sharply raked bows. On boats like this, you may want to tie the fore and aft slings together to prevent slippage.

DOCK MANEUVERING

As mentioned earlier, the first thing to do after launching is to tie your boat up to a nearby dock. One tip to keep in mind is that given an option, it's easier on your boat if you dock into the wind or current. When you tie up on the *leeward* (or downwind) side, the wind or current pushes the boat away from the dock. Docking on the other side enables the wind to push the boat into the dock and causes unnecessary wear and tear.

An easy way to tie your boat up is to use a figure-eight *cleating knot*. Take the end of the line and run it completely around the base of the *cleat*. Then come up and over the near horn. Cross over the cleat on the way to the far horn and go down and around it. On the way back to the near horn, crisscross the line. Finally, finish it off with a half hitch and pull taut. It's one of the easiest and most effective knots in boating.

CLEATING A LINE

Start by looping the line around the base of the cleat. This will help relieve some of the stress on the knot.

Cross the line over the cleat and loop around one of the cleat horns.

After rounding the horn, crisscross the line over the cleat.

Make a loop in the line, twist it, and place it over the horn.

Now pull it tight. It's quicker and easier to tie than to explain.

Pulling into a slip or dock space requires a bit of boat-driving skill, even at idle speed. The first thing you'll notice is that a boat, unlike a car, steers from the stern. When you turn the steering wheel, the stern moves rather than the bow. The second thing you'll notice is that a moving boat, especially if it's equipped with outboard or stern-drive power, only responds to steering input when the engine is in gear. In neutral, the steering wheel is mostly useless. For this reason, a good skipper will use a combination of steering adjustments and small bursts of power to glide toward a dock.

The best method for docking is to set up a straightline approach at about a 30-degree angle. As the bow approaches, turn the wheel away from the dock and then shift into neutral. This action will tend to swing the stern around parallel with the dock. When the boat is a few feet away, turn the steering wheel all the way over in the direction of the dock, shift into reverse, and give the throttle a brief burst. This will do two things: It will stop the boat's forward progress, and it will pull the stern in close for tying up.

After all passengers and supplies are aboard, it's time to pull away from the dock and head out for a day of boating adventure. First, start the engine and make sure it's warmed up and running properly. Then get a deckhand to untie the lines and push the bow away from the dock. Shift into forward and pull away from the dock at idle speed. Make sure you don't turn the wheel too sharply or you will send the stern banging into the dock. Take a narrow departure angle and hold it until you are safely away from the dock. Then you can turn as sharply as you like.

STERN STEERING

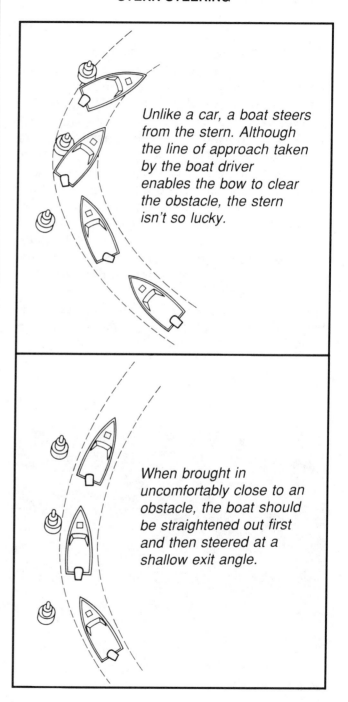

Unlike a car, a boat steers from the stern. Although the line of approach taken by the boat driver enables the bow to clear the obstacle, the stern isn't so lucky.

When brought in uncomfortably close to an obstacle, the boat should be straightened out first and then steered at a shallow exit angle.

ON THE WATER

PULLING INTO DOCK

The best way to pull into a dock is to idle in slowly at an angle.

When you approach the dock closely, shift the throttle into neutral.

While in neutral, turn the steering wheel in the direction of the dock as far as it will go and shift into reverse.

With the steering wheel turned toward the dock and the throttle in reverse the prop will complete the maneuver by pulling the stern into the dock.

Adjusting to the wind or current is as important in leaving a dock as it is in approaching one. If you're heading out into the wind or current, you may have a hard time getting the bow out far enough to safely pull away. In these conditions, your best bet is to put a fender between the stern of your boat and the dock. Then untie all the dock lines except for one, a *spring line* that's cleated amidships on the dock and leading to a cleat on the stern. Now you're ready to get under way.

Turn the wheel all the way toward the dock and shift into reverse. Use enough throttle to pivot the bow well into the wind or current. Then shift into neutral and quickly untie the spring line. With the bow pointed well away from the dock, shift into forward and head safely out of the marina. Remember not to turn too sharply until the boat's well away from the dock.

PULLING AWAY FROM DOCK

With little wind or current to worry about, the first steps are to untie all lines and push off from the bow.

Now properly angled, idle away from the dock. It's a good idea to go a safe distance before turning the wheel sharply.

ON THE WATER

Then turn the steering wheel toward the dock and shift into reverse. This maneuver will swing the bow free of the dock. Now untie the line, shift into forward gear and pull away. Make sure to use a fender at the stern of the boat when the dock is without padding.

When wind or current makes getting the bow away from the dock a problem, start by tying a line to a stern cleat as shown above.

BOAT HANDLING

Numerous books have been written on the subject of boat handling. A word of advice: Read one. There's no substitute for being well versed in this complex subject. Even better, take one of the many boating courses offered around the country.

In addition to the number of skills already discussed in this chapter, boaters need to know the rules of the road regarding rights of way on the water. They need to know how to cope with hazards, identify marker buoys, tie knots, and handle emergencies. They also need to be familiar with basic troubleshooting, towing procedures, navigation, anchoring, heavy-weather seamanship, and much more.

Two of the best and most comprehensive books available on the subject of boating are *Chapman Piloting: Seamanship & Small Boat Handling* by Elbert S. Maloney, and *Stapleton's Powerboat Bible* by Sid Stapleton. Both books are one-stop reference books published by Hearst Marine Books. For free operating information visit the local district headquarters of the U.S. Coast Guard or call (800)368-5647, or contact the Boat/U.S. Foundation at (800)336-2628.

Organizations that offer excellent boat safety and handling courses are the United States Coast Guard Auxiliary, Washington, D.C. 20593, which can be reached by contacting a local Coast Guard district headquarters or by calling (800)368-5647; and the United States Power Squadrons, Raleigh, N.C. (919)821-0281. Classes offered by these volunteer organizations may require a minimal fee.

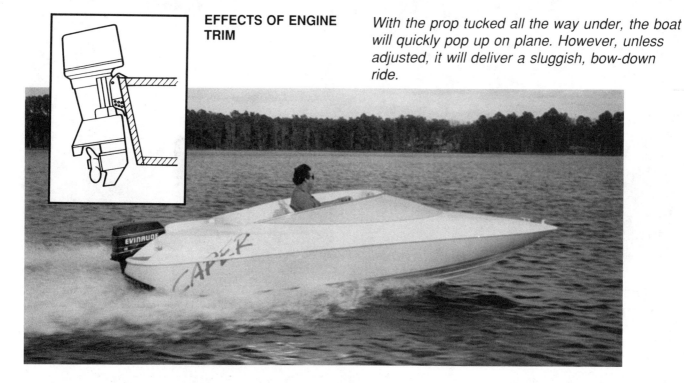

EFFECTS OF ENGINE TRIM

With the prop tucked all the way under, the boat will quickly pop up on plane. However, unless adjusted, it will deliver a sluggish, bow-down ride.

When the prop is correctly trimmed, the boat rides with a level attitude and the bow stays efficiently out of the water.

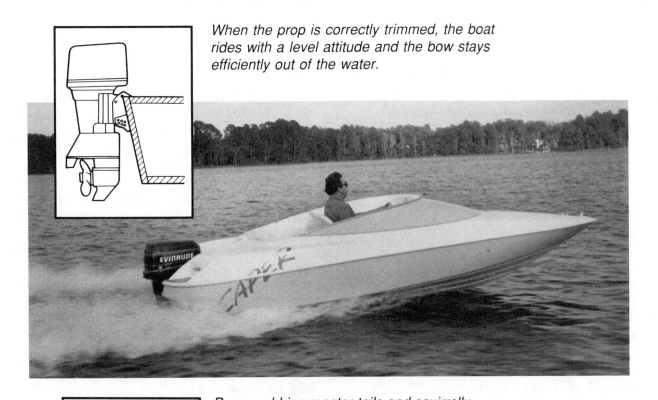

Power-robbing rooster tails and squirrelly handling are the result of excessively high prop trimming. Engine damage may also occur if insufficient water is fed to the pickups.

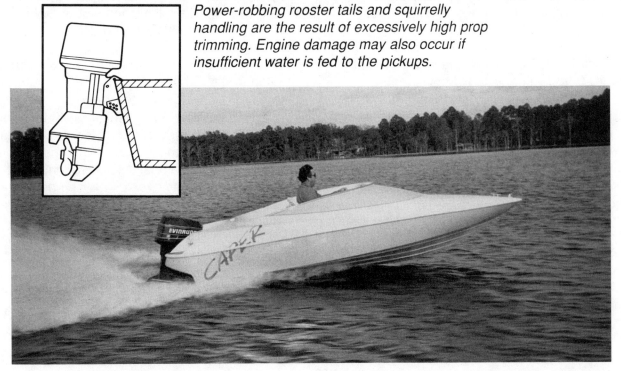

HAULING OUT

As expected, the correct procedure for hauling out a boat is basically the reverse of what it takes to launch it. Pull up to the dock and let all ashore who are going ashore, especially the tow-vehicle driver. Then, either tie up to the dock or find a place to idle and wait until the tow-vehicle driver backs the trailer into the water.

Two tips for the tow-vehicle driver: 1. Make sure the trailer backs straight into the water, otherwise the boat may not be able to correctly position itself in the cradle; and 2. Make sure the waterline intersects the trailer wheels at roughly the same point during launching.

Now, the tow-vehicle driver engages the emergency brake, shifts the transmission to park to help hold the vehicle in place, and signals the boat driver to pull onto the trailer. The boat driver shifts into forward gear, trims up the prop to prevent striking the bottom, and idles toward the trailer. It's important that the boat driver approach on a straight line and make as few steering corrections as possible. If the boat is heading for an off-center approach, the best bet is to start over again.

When the bow comes to rest on the trailer bed, the tow-vehicle driver must evaluate whether or not the boat is properly seated on the bunks or rollers. If it isn't seated properly, then the boat must be backed off the trailer and pulled in again. No boat should be towed when it's misaligned in the cradle. Now, the tow-vehicle driver must determine how close the bow is to the winch stand. If it's nudged up to the bow stop, then the driver may simply hook up the winch line and crank the winch handle a few times to snug it in place. If the bow is several inches away from the winch stand, then the boat driver must shift into forward and bump the throttle to move the boat closer. When the proper position is achieved the tow-vehicle driver hooks up the winch line and secures the boat to the trailer.

Once the boat is securely in place, the boat driver should turn off the engine and tilt the lower unit completely out of the water. The final step is to attach the winch-stand safety chain. Remember that the weight of the boat alone isn't enough to hold it in the trailer bed. Many boats have been dropped on the launch ramp because they weren't secured. Also, now is a good time to crank the winch tight and flip the safety latch.

Now, the tow-vehicle driver should shift into low gear and pull slowly but firmly up the ramp. If the marina has a wash-down area, this should be the boater's first stop. Remove the drain plugs to let out the bilge, and then proceed to give the boat and trailer a thorough spraying. Flush out the engine's cooling system, too, if it has a convenient hookup or you've brought a flushing device. Finally, take one more look at the boat's alignment on the trailer. If it's misaligned, you may have to drop it back in the water to properly seat it in the trailer bed.

When hauling out by sling or lift, it's important to keep in mind that wet boats are slippery. For this reason, it's a good idea to tie the slings or straps together for extra security. If the boat doesn't settle properly in the trailer, don't be embarrassed to ask the operator to reposition it. Never drive off with a boat that's misaligned on the trailer bed. Then head over to the wash-down area to spray off saltwater or grime.

After a long day on the water, it's tempting to hurry through the final procedures for stowing gear, covering the boat, and tying it down. Certainly there's less joy in leaving the marina than in arriving, but you won't savor the experience unless all trailerboating responsibilities are properly and safely executed.

ON THE WATER

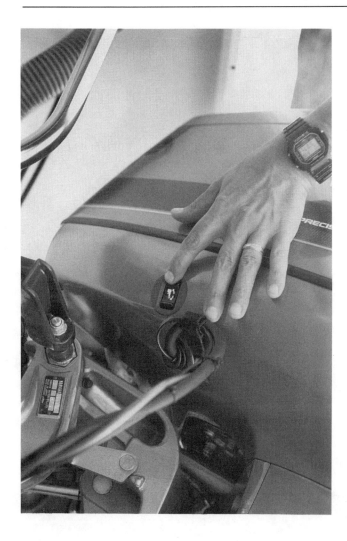

The work of trimming up the engine once it is loaded onto the trailer has been eased by the installation of built-in trim buttons on many new outboard and stern-drive engines.

Checking that the boat safety chain is hooked to the bowline is an important security step before hitting the road.

Before hitting the road after hauling out, crank the winch tight and flip on the safety latch.

CHAPTER 7

MAINTENANCE

For many of us, the era of climbing under the family vehicle or poking extensively beneath the hood is pretty much over. Certain small jobs often can still be performed, but for the most part modern cars and trucks—equipped with microprocessor brain boxes, fuel-injection systems, and an undercarriage that's crisscrossed with stabilizers, A-arms, and torsion bars—are now well beyond the maintenance capability of most weekend mechanics.

This isn't necessarily the case with trailers, tow hitches, boats, and marine engines. In fact, frequent inspection and hands-on maintenance are the norm for the majority of trailerboaters. Not only will these tasks save you money and ensure that everything is in top operating condition, they'll guarantee that your equipment preserves its value for future resale.

There are many ways to approach trailer rig maintenance, and a good one is to start at the ground, with the tires and wheels, and then work up to the trailer, the hitch, and the boat.

TIRES AND WHEELS

The first thing to do here is to check the maximum capacity rating for your trailer tires. This information is stamped on the sidewalls. Multiply the rating figure by the number of tires on the trailer, and then make sure the total is equal to or greater than the load. Maintaining a safety margin of several hundred pounds is recommended.

Next, check the inflation pressure on the tires and make sure it, too, complies with stamped sidewall information. This inspection should be done frequently, because a fully loaded trailer should never run with underinflated tires. Underinflation causes overheating, which leads to premature wear and blowouts. Overinflation causes the tread to disintegrate.

Now it's time to inspect each tire, including the spare, for signs of excessive wear, both in the tread area and in the sidewalls. Although few trailer tires accumulate large amounts of highway miles, it's worthwhile to measure the tread and make sure the depth is not much less than a quarter of an inch. If tread depth is approaching an eighth of an inch, or if the built-in wear indicator begins to show through, it's time for replacement. On the sidewalls, check closely for signs of cracking. Long-term exposure to the sun's ultraviolet rays deteriorates tires and causes stress cracks to form in the sidewalls. These cracks are a warning sign that the tires may fail when subjected to a shock load.

When checking wear, don't forget to pay special attention to unusual tread-wear patterns, such as excessive wear on one side or flat spots. These problems may be caused by misaligned or unbalanced wheels, a misaligned axle, or a poorly functioning suspension.

Before moving on, make sure that all the wheels have a full complement of lug nuts and all are wrench tight. Now, check the wheel rims for signs of rust and dents. If there are rust spots, scrape them with a wire brush and touch them up with anticorrosion paint. If the rim has a dent that appears to interfere with proper tire seating, the best bet is to replace it. Again, don't forget to check the spare. Most of the above duties are recommended for every trip.

WHEEL BEARINGS

Without wheel bearing protectors, trailerboaters are required to dismantle the hubs and repack the bearings with grease every time the wheels are immersed. For most trailerboaters this would mean doing it every time the boat is used. Fortunately, most modern trailers are equipped with bearing protectors, which greatly reduce maintenance. (Refer to the section on wheel bearings in Chapter 2.) If your trailer doesn't have bearing protectors, the best advice is to buy and install them, an investment that will save hours of maintenance duty.

The first thing to look for in routine bearing maintenance is to see if the grease seals are holding. A small oil film is normal around the seal area, but it shouldn't be excessive. If it is, then the seal may be worn. Replacement of a bad seal requires removing the complete hub assembly. Pay special attention to seals on brake axles. Oil that leaks onto the brake drums adversely affects the linings.

For routine bearing-protector maintenance, press the edge of the spring-loaded piston inside the cap to see if it will move or rock. If it will, then it's still filled with grease and no action is required. If it doesn't rock, then it's fully depressed and needs grease. To refill the bearing protector with grease, push the end of a cartridge-type grease gun over the bearing protector's refill fitting, which is located in the center of the cap. Then, squeeze until the spring is fully extended. This inspection should be performed during every trip, if needed. One final tip: To keep off dirt and grime, buy a plastic cover to fit over any bearing protector that doesn't already have one.

MAINTENANCE

Handy grease fittings on bearing protectors make it easy work to lubricate wheel bearings after immersion in water.

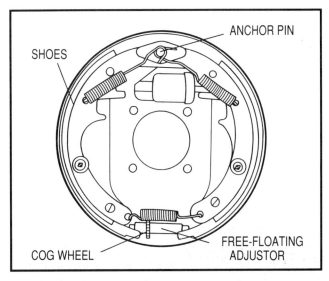

Unlike automotive brakes, trailer brakes aren't self-adjusting. Improper adjustment is a primary cause of brake failure, so they should be checked regularly.

BRAKES

Trailer brakes, especially surge brakes, are not self-adjusting. Fortunately, the adjusting process is fairly simple. First, raise the wheel off the ground. Then, remove the dust grommet from the adjusting slot. It's located on the lower part of the back side of the brake assembly. Now, insert a brake-adjusting tool into the slot and move it toward the top of the drum. This action rotates the adjustment cog. At this point, rotate it as far as it will go, so that the brake shoes inside the drum are fully tightened. Now, back off the shoes by reversing the rotation of the cog. This should be done to exact specifications found in your owner's manual. Backing off five to ten notches, or when the wheel begins to turn freely, is a typical recommendation.

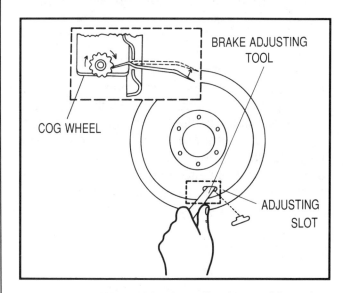

Insert the dedicated brake-adjusting tool into the adjusting slot found on the bottom of the back side of the brake plate. Rotate the cog wheel by moving the handle of the adjusting tool in an up and down motion.

After replacing the grommet, move on to the remaining trailer wheels. Make sure all brakes are set to the same adjustment point. Then examine shoes and lining for signs of wear. Replace as necessary. This procedure should be performed annually. If you're not comfortable performing this or any technical task in this section, the best approach is to go to a good service shop.

SUSPENSION SYSTEM

A few light duties are required to keep the suspension system in top condition. If your trailer is equipped with coil springs, torsion bars, or shock absorbers, the maintenance routine consists of inspecting rubber bushings, if any, to see if they're dried out or cracked or show other signs of wear. Replace if necessary.

For leaf-spring suspensions, the end pivot points should be lubricated and all rubber bushings must be checked. Then make sure the steel outer surfaces of the leaves themselves aren't showing signs of serious wear. Finally, check the leaf attachment points—the hangers, shackles, equalizer bars, leaf clips, and all attendant bolts. Replace worn parts if necessary, and make sure all screws and nuts are wrench tight. Once-a-year attention to the suspension system should suffice.

FRAME CARE

Rust is a constant worry for those who don't own galvanized-steel or aluminum trailers. For these boaters, the best advice is to keep the trailer away from water as much as possible, and if used in saltwater, rinse it off immediately after each immersion. Clean off any dirt buildup as soon as possible, since these deposits can trap moisture, and be sure to clean out deposits that work their way into open-end tubing. For painted trailers, touch up scrapes immediately and apply a wax or sealer once a year. This should be done after washing it off, removing loose rust, and scraping the surface with a wire brush. Finally, give galvanized trailers an application of corrosion-resistant coating once per season.

The next important worry for boaters is structural soundness. With the boat off the trailer, check all roller assemblies, bunks, and other frame components to see if anything looks bent or pushed out of alignment. Shake and wiggle everything to check for sturdiness. Look for cracks, bends, or other signs of wear, especially on rubber roller bearings or keel rollers. Replace as necessary. Carefully inspect welds, and make sure all screws and nuts are wrench tight. Except for washing down, these duties need only be done annually.

LIGHTS AND WIRING

The biggest enemy of the trailer's electrical system is oxidation on the contact points. Caused by moisture, oxidation can result in rust or other deposits that interfere with solid electrical contacts. The best way to combat this problem is to smear a light grease on all plug prongs, receptacles, light-bulb sockets, and wire connections.

A good tip for the wiring harness plug is to scrape off all surface deposits on the prongs and in the receptacle holes, and then dab both with grease. This is especially important when storing for a long layover, but it's also a good idea for

MAINTENANCE

permanent use. On long trips, after dabbing the plug and receptacle with grease and connecting them, it's a good idea to wrap with electrical tape to reinforce the seal.

Just as wheel bearings heat up after a long drive, so do bulbs in the trailer's lighting system. Maintenance problems associated with immersing hot bulbs in cold water can be avoided by letting the trailer cool down for at least fifteen minutes after a long ride or installing waterproof lights. It also helps to disconnect the wiring-harness plug so that the lights remain off when submerged.

Finally, make sure all connections are tightly secured, and no fraying or pinching is visible. Constant vigilance is always the best policy, but a thorough inspection of the electrical system need only be done once a year.

TONGUE COMPONENTS

First, check the coupler for rust, dents, cracks, and other signs of wear or stress. Make sure the coupler socket has maintained its shape, the ball clamp is tightly adjusted, and all screws and nuts are wrench tight. It's also a good idea to give all pivot points and moving parts a shot of lubricating oil.

Then open the surge-brake housing and check the level of braking fluid. Refill as necessary. Lubricate pivot points, and make sure the emergency lanyard shows no signs of fraying.

Next, check the safety chains on both the tongue and winch stand. Also, check the bow stop and winch line. All should be closely inspected for signs of wear and replaced when necessary. Make sure all screws and nuts are wrench tight on the winch, and lubricate moving parts for long-term protection. None of these components should ever be left unchecked for long, but detailed inspections should be done at least annually.

THE HITCH

Inspect the hitch ball for cracks and flat spots. Be sure it's wrench tight. Then, grab the hitch and firmly shake in all directions to make sure it's solidly connected.

Now, climb beneath the tow vehicle and visually check the condition of the connections and welds. Make sure the hitch cross bars and connection points show no signs of serious stress, flex, or wear. Finally, check the mounting bolt holes to see if they're becoming elongated through stress. Again, constant monitoring is required for safety, but a formal inspection need only be done annually.

THE BOAT

As mentioned before, washing down the hull, decks, and cockpit should take place after every use, especially when the boat is used in saltwater. Check the exposed surfaces for marine growth and other stains and clean them with a biodegradable hull cleaner before they become embedded. Finish with a coat of marine polish to seal the gel coat.

While you're cleaning the hull, check it for dings and cracks, and blisters or bulges, which may be signs of water migrating through the gel coat. Repairing these potentially serious problems with a fiberglass repair kit isn't as hard as you might

To keep water damage from spreading through the hull, clean out gouges and cracks, add epoxy-based fiberglass filler, allow to harden, and then sand to a smooth finish before adding gel coat.

think. Carefully drain and dry the damaged area, then patch it with an epoxy-based fiberglass filler. Wait until it sets and then sand to a smooth finish. Owners of aluminum boats can bang out dings and dents with a rubber mallet, but tears and gashes may require welding.

If your boat has teak trim, clean it regularly with a nonacid cleaner. Then apply teak brightener and teak oil sealer. Padded panels and upholstered seats can be cleaned with a good vinyl cleaner and restorer.

One final tip: After every trip it's a good idea to drain, clean, and dry your marine sanitation device (MSD).

THE ENGINE

Not only are there inboard, outboard, and stern-drive engines on the market today, but there are four-cycle outboards, two-cycle inboards, and a growing number of electronic fuel-injected models. Consequently, the most important maintenance tip for any trailerboater to keep in mind is to read the owner's manual thoroughly. A number of warnings and liabilities are involved with marine engines, and you should be fully aware of them.

That said, there are a number of things you can do to keep your engine running in top shape. The first is to wash down the exposed drive unit after every use. Also, it's a good idea to attach a water hose to a *flushing device* on an outboard or an I/O and completely flush out the raw-water cooling system. Again, these procedures are especially important after use in saltwater.

Now, check the prop, skeg, and cavitation plate for dings and other signs of serious wear. Minor dings can be repaired by banging them out and filing jagged edges. Damage to a prop blade can mean damage to the prop's shock-absorber bushing, and may require replacement.

Now move up to the engine and make sure all wires, electrical connections, and clamps are secure and in good condition. Also, check the condition of the hoses for brittleness or cracking. On inboard and stern-drive engines, check the condition and tension of all drive belts. Repair or replace as necessary. Finally, make sure all fluids— engine oil, gearcase lubricant, power trim/tilt fluid, transmission fluid, battery water, and hydraulic brake fluid—are at their proper levels.

There's quite a bit more maintenance that you need to do to your engine, boat, trailer, and hitch, and this will be covered in the following section on long-term storage.

MAINTENANCE

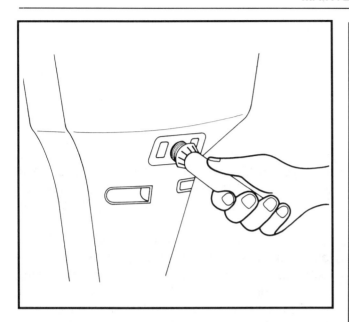

Flushing the engine with freshwater has been made easy on many new motors with the addition of a built-in flusher coupling that's threaded to fit a garden hose.

LONG-TERM STORAGE: WINTERIZING

Hibernation, or any long layover, can be just as destructive to a boat/trailer rig in the Deep South as to one in the deep freeze. Inactivity is the culprit, and if you don't prepare for it, you run the risk of incurring costly repair bills, devaluing your investment, and shortening the life of your rig. Owner's manuals generally recommend that you take your rig to a service shop for prepping and a tune-up before long-term storage. However, many boaters spend an afternoon in the fall to do it themselves. Starting with the trailer and moving up, here are some tips to make sure that when you put your rig to bed it wakes up as good as new.

Check the condition of the tires and rims as recommended earlier. One good tip to help guard against premature rusting is to remove each wheel and thoroughly scrape the rims with a wire brush. Then, apply a complete coat of paint or an anticorrosion coating. Make sure to coat the rim right up to the tire bead.

In addition to making sure the bearing protectors are full of grease, conventional wisdom calls for trailer bearings to be disassembled and repacked with grease about every two thousand miles. The best time to do this is before a long layover. The messy job involves jacking up the trailer, removing the bearing protector, removing the wheel and hub, removing the inner and outer bearing assemblies to expose the bearing rings, replacing worn parts, cleaning all components with solvents, and then packing the bearings with grease before reassembling. Note that the rear bearing seal will most likely be damaged during disassembly, so be prepared to install a new one before you start. Plan on spending at least one hour per wheel to perform the job correctly. However, if you routinely use a good quality lubricant and maintain proper levels through the season, you may be able to skip this duty every other year, especially if the grease doesn't appear to be contaminated or broken down during a visual inspection.

As mentioned earlier, manual brake adjustment should be done once a year. The same is true for checking the condition of brake shoes and linings. While disassembled, scrape out all rust and replace worn parts as necessary. In addition, now is a good time to check that the hydraulic fittings are secure. If you find any leaks, or if the brake-fluid level in the reservoir is lower than the master cylinder ports, or if any component of the brake's hydraulic system has been disconnected during regular maintenance, the system must be bled.

To bleed hydraulic brakes, connect a rubber hose to the bleeder fitting and submerge the free end in a container holding some new brake fluid. By submerging the hose you ensure that no air is returned to the system. Then loosen the bleeder screw and pull the breakaway lanyard on the surge actuator. Make sure the master cylinder is filled during this procedure. Bleeding is completed when expelled brake fluid is free of air bubbles. To finish up, close the bleeder screw securely, remove the hose, and refill the master cylinder.

Thorough attention must be paid to the frame, bunks, rollers, suspension and electrical systems, tongue components, and the hitch prior to layover. Care of these parts is covered earlier in this chapter. Remove the hitch ball and store it indoors for complete protection. One additional layover tip is to spray moisture-dispersing oil on the hitch assembly, winch, leaf springs, and undercarriage assemblies.

Your boat should be stored in absolutely clean condition, since all dirt and stains will become embedded during storage. As mentioned earlier, use a biodegradable hull cleaner and then finish with a coat of marine polish to seal the gel coat. Repair all blisters, water bulges, dings, and cracks as necessary, and perform touch-up work on dock rash and other bruises. Also, check all deck hardware for signs of wear. Repair as necessary.

If your boat spends a good portion of the summer season in a dockside slip, now is the time to apply a good coat of antifouling paint. Finally, remove the drain plugs and flush out the bilge area, livewells, baitwells, and freshwater system. Drain all thoroughly. Before covering, don't forget to open all hatches and doors.

Teak trim, upholstery, and the marine sanitation device (MSD) require the attention mentioned earlier. If your boat is equipped with a built-in MSD or a shower, the system must be completely flushed out, drained, and then filled with a water-system antifreeze. Now is also a good time to vacuum all carpeting.

Wash the outdrive portion of your engine and flush the raw-water cooling system as indicated above. Drain and dry thoroughly. Check the prop, cavitation plate, and skeg as recommended earlier. Remove the prop and store it in a locking

BLEEDING HYDRAULIC BRAKE SYSTEM

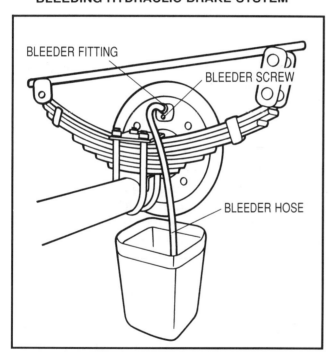

Whenever a hydraulic component of the brake system has been disconnected or when the level of fluid in the master brake cylinder has dropped below the level of the delivery port, the brakes must be bled until the fluid is free of air bubbles.

MAINTENANCE

Removal of the prop before running the engine with freshwater flushers is made easy with a dedicated prop wrench.

compartment in the boat. Finish up the prop shaft by spraying the splines and exposed components with a light grease for corrosion protection.

Now, remove the oil fill and vent plug from the lower unit and drain the oil. Inspect for metal filings, a milky color, or a burnt appearance. These signs indicate problems that need further attention by your service shop. If all looks fine, inject gearcase lubricant into the lower fill hole until lube appears at the top port. Inspect the zinc anodes for wear. Carefully placed zinc anodes are used to prevent natural electrolysis that occurs when dissimilar metals are immersed in water. If they appear to be reduced in size, replace them.

Owners of inboard-equipped boats need to look beneath the hull and inspect the drive-shaft packing, propeller, and rudder. Tightening the packing should only be done when the boat's in the water, so this becomes a job for the spring. Check the condition of the rudder and steering assembly and make sure all screws and nuts are wrench tight. Remove the prop and store it. Spray the prop shaft with a water-dispersing oil.

Before moving to the engine, lubricate the drive shaft, engine coupler spline, U-joint, hinge pins, pinion gears, swivel pin, gimbal bearing, steering-cable ram, and trim/tilt assembly using greases that are specifically formulated for each area. This is also the time to fill your power tilt/trim reservoir, and to disconnect the speedometer hose connection to drain water out of the speedo pickup. Use touch-up paint on exposed metal surfaces to halt the onset of rust, but make sure you don't coat the zinc anodes and render them useless. Outboards and I/Os should be stored in the full trim-down position to prevent strain on the hydraulic cylinder seals.

After performing the engine inspection suggested earlier regarding wires, electrical connections, hoses, clamps, and belts (note that tension should be loosened during long-term storage), it's time to clean and examine the flame arrestor (on inboards and I/Os), fuel-line screen, fuel filter, and oil filter. Replace as necessary.

Top off the fuel tank and add fuel stabilizer to prevent internal gumming as the fuel slowly breaks down during storage into harmful resins. You can also add a fuel water absorber to your topped-off tank to remove moisture that forms through condensation, and then pour in an additive that coats internal engine parts with corrosion inhibitors.

Change the engine oil by draining it while the engine is still warm. To fully remove all remnants of old oil in I/Os and inboards, use a hand pump hooked up through the dipstick hole. Top off the oil reservoir and add oil stabilizer to prevent viscosity changes. When replacing the oil filter, make

Don't underestimate the importance of covering a boat. During layovers a cover protects the upholstery and carpeting from sun and rain. On the road it helps improve fuel economy.

---

sure to prefill it with oil so that the engine has an immediate supply of lubricant upon starting.

Now run the engine briefly to disperse the protectants throughout the engine systems. Remember that you need to use a flushing system to keep water in the cooling system whenever you run a marine engine on land. Also, keep the transmission in neutral and remove the prop to avoid the danger of spinning blades.

Just before the final engine shutdown, spray fogging oil through the carburetor intake. This should be done after the fuel line has been disconnected and the engine is about to burn up its available supply of fuel. After the engine cools down, remove the old spark plugs and spray fogging oil into the cylinders. Then crank the engine with one or two quick bursts to spread the oil on the cylinder walls. Install new, correctly gapped plugs if necessary.

Finally, spray all exposed engine assemblies and surfaces with moisture-dispersing oil, and

MAINTENANCE

then top off power steering, transmission (for inboards), and battery fluids. Outboard motors will completely self-drain of coolant when you trim them all the way down, but inboards and I/Os require removal of drain plugs on the engine block. Make sure all water is drained from the exhaust manifolds, exhaust risers, oil cooler, and water pump. An increasingly recommended alternative to draining is to pump environmentally approved antifreeze through the raw-water pickup. However, if you completely drain your coolant system, replace the water pump impeller when the boat is put back in service in the spring.

Don't forget to check the condition of your battery cables and terminals. Scrape off all deposits and dab with grease for protection. To maintain battery life, it's a good idea to remove it and store in a cool, dry location.

Now you're ready to close up the boat under a canvas or plastic cover. First, remove all pieces of electronic equipment, trolling motors, and so forth. To protect the terminal ends, spray them with a little silicone and cover with tape to prevent corrosion. On inboards only, plug the exhaust pipes to keep out moisture.

Then, bring in all canvas or fabric-covered items that are removable and wash off mildew, if any. Also, spray zippers and snaps with a light lubricant to prevent rust. To prevent mildew during storage, coat canvas, plastic, upholstery, and carpeting with a mildew-control spray. To make sure the boat covering is truly waterproof, coat it with a waterproofing fabric treatment. If your canvas cover comes with a center pole to promote water runoff, make sure to use it. Don't forget to tie down the cover securely, to leave ventilation holes, and to place pads beneath lines and other points that may chafe the hull.

To prevent movement during the layover, put chocks beneath the wheels or remove the wheels and set the axles on blocks. Blocks also belong beneath the trailer tongue and the back end to keep the rig level, but slightly bow up to let water drain. Finally, don't forget to visit your rig occasionally to check on its hibernation.

In the spring, your engine should start right up after the battery is charged. First give the engine and drive a complete visual inspection, and top off the fuel tank with some fresh fuel. One good tip to remember is to hook up a flushing device and run the engine to burn out the protectants before heading out. It will save you a potentially embarrassing surprise on the water.

CHAPTER 8

ACCESSORIES

Just when you think you've got your trailer and boat fully outfitted and geared up, along comes some tempting new equipment that you simply have to have. While most of this gear falls into the optional category, some of it is virtually essential for safe and secure trailerboating.

Equipment in this latter category consists of accessories and tools you'll need for routine road and water operation, unexpected emergencies, and ongoing and seasonal maintenance. Another important category of essential gear is locks and alarms. Many boaters don't realize it, but police and insurance statistics indicate that the prime target for bandits is a trailerable boat that's eighteen to twenty-six feet long. There's a lot you can do to make sure bandits don't make off with your valuable possessions.

Tools, accessories, and gear are a vital part of trailerboating, so here's a look at some items that will help ensure the safe and secure operation of your tow vehicle, hitch, boat, motor, and trailer.

GEARING UP THE BOAT

The most important item you need for your boat is a fully outfitted *Coast Guard safety package*. This is a legally required set of gear that begins with *personal flotation devices (PFDs)*—one for each person aboard and a throwable float for boats longer than sixteen feet. For boats less than sixteen feet long, you must carry either one PFD or a throwable float for every person aboard. If your boat is longer than sixteen feet, you must also carry a visual distress signal. Hand-held or rocket flares satisfy this requirement. Finally, all boats

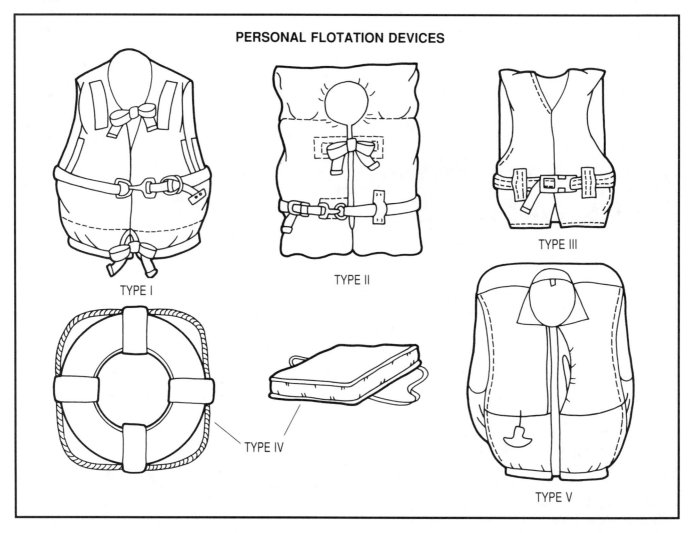

PERSONAL FLOTATION DEVICES

As part of the required Coast Guard safety package for boats there must be at least one PFD on board per passenger plus one throwable.

longer than sixteen feet except nonpowered open vessels must carry a fire extinguisher and a horn, whistle, or bell.

You can either spread out the Coast Guard safety gear among several stowage compartments or keep it all together in a bag for easy accessibility. Whichever method you use, read the labels on the fire extinguisher and flares at the start of each season, and replace them before the expiration date.

ACCESSORIES

All boats must be properly registered and display their identification numbers. Position is important, too. The numbers must be well forward and close to the bow.

Carrying an anchor isn't a Coast Guard requirement, but it's highly recommended. In addition to enabling you to tie up in a desired location, an anchor can be helpful in emergencies. Make sure you get sturdy anchor line and a chain or rode if you intend to use it frequently. Also, make sure the length of line is sufficient for the size of the boat and the depth of the water.

For boaters who launch in the ocean or on large lakes and rivers, a marine VHF radio is virtually an essential piece of equipment. Either a permanently installed unit or a portable hand-held model should accompany boats that go well offshore on big water, and make sure it's licensed.

A canvas, cotton, or polyethylene boat cover is also an important accessory. Not only will it protect your boat from the elements—both the destructive summer sun and the harsh winter cold—but it will also prevent your rig from becoming a wind trap while riding down the highway. A properly tied-down boat cover will actually help improve gas mileage on your tow vehicle during long road trips.

When buying a cover, make sure you premeasure your boat and buy one that's a good fit. You will encounter problems if your cover is either too large or too small. Fortunately, many cover manufacturers pattern their products to builder specifications.

When it comes time to put your rig to bed for long-term storage, your boat cover should be prepared to deal with harsh conditions. Chief among these is the stress of sagging that results from puddling rainwater. To prevent this situation from turning into a destructive leak, you can create a support structure beneath the cover to promote runoff and reduce sagging. One method is to use lengths of nylon web strap to run from the bow to the transom. By running up and over the windscreen, the straps make sure that the cover doesn't sag. Another method involves the use of telescoping rods to create a self-supporting structure in the cockpit of the boat and in the bow, if necessary.

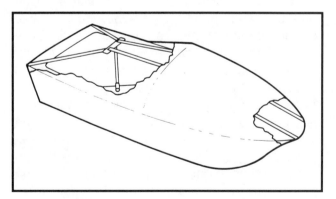

One method to protect your cover from the stress of puddling rain water is to run supporting straps beneath it and set up a pole in the cockpit.

In addition to the rope that's used to pull the boat cover tight, it's a good idea to loosely run several lengths of line on top of the cover for extra wind protection. One line each should go over the bow, the stern, and amidships. Special nylon web straps with hook ends and ratchet buckles are built for this purpose. However, sturdy rope will work just fine for long-term storage. For wind protection during high-speed trailering, web straps are better.

TRAILER ACCESSORIES

All trailers should be equipped with a tongue jack. If yours isn't, or if yours is a simple drop jack with a steel foot, you may want to upgrade to a swivel jack fitted with a dolly wheel. The dolly wheel enables you to easily move the trailer around during the hookup process. The swivel mechanism enables the jack to swing up and out of the way. Another benefit with the swivel jack is that the dolly wheel doesn't have to be removed and stored—and possibly lost or forgotten—when not in use.

Trailerboaters may occasionally find it necessary to change a flat tire on a fully loaded trailer, and to do this you need a heavy-duty trailer jack. There's an interesting new jack with no moving parts that performs this function. It's a D-shaped, one-piece unit that is placed beneath the axle: When you drive the trailer forward, the jack grips the road and rotates upward until it stands on end. The simple unit has a two-ton capacity and is so small and lightweight that it's easily stowed.

As mentioned earlier, wheel-bearing protectors are invaluable maintenance-savers, especially if you immerse the trailer wheels during launching or do some of your boating in saltwater. If your trailer isn't equipped with wheel-bearing protectors, it's highly recommended that you get them.

Several brands come with see-through caps for easy lubricant-level inspection. Some have an automatic level indicator that recedes or pops out depending on the amount of grease in the cap. And a few models have pressure-release mechanisms that allow excess lubricant to gradually escape through the rear seal before damage can occur.

A new type of hub system enables you to repack the bearings with lubricant without disassembling the wheel bearings. When it is installed, a special fitting is located in the seal of the rear bearing. By forcing new lubricant into the rear seal, you expel the old grease and replace it with a fresh supply. For simple replenishing, you use the grease fitting in the center of the front-bearing cap as you would a typical unit.

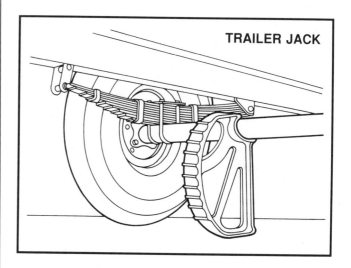

To operate this simple but effective trailer jack you simply drive forward.

ACCESSORIES

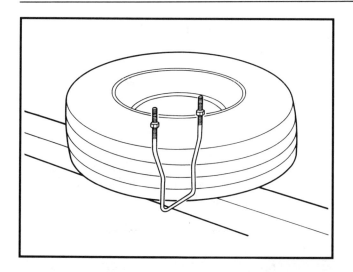
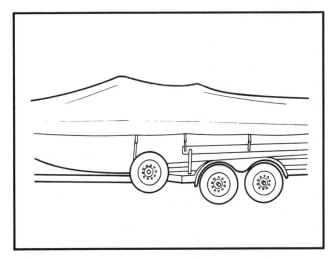

TRAILER TIRE SPARE MOUNTS

Finishing up with the wheel bearings, make sure yours are covered with a double-lip dust cap. You might also consider buying a set of full-wheel hub caps, which not only look good but add a measure of protection.

Smart trailerboaters know that not all service stations stock trailer tires, so they carry a spare. The best place for the spare is on the trailer itself, typically on the tongue. The simplest mounting unit is similar to a hanging U-bolt. It positions the spare tire horizontally beneath the trailer. To position the tire vertically on the trailer tongue or winch stand, you can install a high-mount carrier.

If your trailer doesn't have water-sealed, sub-

Using a high-mount bracket puts the spare tire out of the way next to the hull. A simple low-mount bracket positions it on the trailer tongue.

mersible tail and side lights, it's probably a good idea to install them. Even if you routinely unplug the tow-vehicle/trailer electrical connection before launching to prevent shorting out, cold water and hot lights don't mix. In the long run, submersible light fixtures are maintenance savers.

Tie-downs, similar to the nylon web straps previously mentioned in connection with the boat cover, are important safety elements on the trailer itself. Equipped with hook ends and ratchet buckles to take in slack, tie-downs are commonly used to make sure the boat stays on the trailer while traveling. They run from the trailer frame to several key attachment points—the transom lifting hooks, the bow eye, and even to cleats on the gunwales. They can also be used to secure the battery and gas tank. One tip to remember: If your tie-down comes into contact with the boat, insert a pad to protect the gel coat from chafing.

Other helpful trailer accessories include: a power winch, for heavy boats; a tongue walk ramp to provide sure footing during launching and retrieval; and trailer guides, to make it easy to position your boat on the trailer when hauling out.

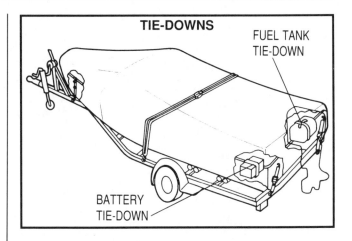

Tying a boat down securely to a trailer means throwing a strap not only across the bow, but also across the battery, the gunwale, the fuel tank, and transom.

TOW VEHICLE EQUIPMENT

Probably the most important accessory for your tow vehicle is a set of large, extended mirrors. By law in most states, you must have mirrors on both sides of the vehicle that extend beyond the width of the trailered load. To improve sight lines on the right side of your rig, a stick-on convex mirror is highly recommended.

Two other useful accessories are mud flaps, to prevent road debris from damaging the hull, and a hitch-ball cover, to ward off rust when not trailering.

Two less common accessories that trailerboaters might find useful are hitch guides, which enable drivers to perform solo hookups, and a unique universal hitch that's equipped with three different-size hitch balls.

The most familiar hitch guide design on the mar-

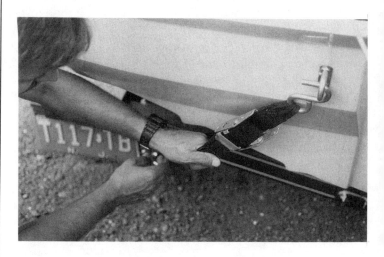

Ratchet-buckle tie-down straps easily and efficiently secure the boat to the trailer by fastening to transom hooks.

ket consists of a base plate that's permanently installed on the back bumper. Two removable plates then fit into the base and form a capture area to guide the coupler to the hitch ball.

A more high-tech system that is common in the world of recreational vehicles is based on electronic radar guidance principles. Two receivers installed on the rear bumper of the tow vehicle receive signals from a removable transmitter that's placed on the trailer tongue. A display monitor on the tow-vehicle dash indicates the location of the coupler in relation to the hitch ball. When the signals merge together, the coupler and hitch ball are properly aligned. Generally speaking, most trailerboaters do without these toys by getting a friend or family member to stand at the rear of the vehicle and give hand signals.

The other unique product of interest to serious towers is a new multiball hitch platform that inserts into a standard hitch receiver box. It's fitted with three different hitch balls (1⅞ in., 2 in., and 2⁵⁄₁₆ in.) and allows towers to go from one size to the other by simply rotating the unit. For trailerboaters who also tow recreational vehicles or horse trailers, this three-in-one unit eliminates the need to carry multiple receiver-type hitch platforms.

MARINE ENGINE GEAR

The foremost accessory a trailerboater can buy for an outboard or stern-drive engine is a motor saver that helps support the lower unit during highway travel. For outboards, a hard rubber or aluminum brace is commonly used. It has a padded V-shaped yoke that's placed beneath the outboard gearcase. The other end attaches to the trailer. The brace or bracket acts as an absorber of shock loads and relieves the stresses of downward force on the transom and trim/tilt seals. To accomplish this benefit for stern-drive engines, two nylon web straps can be run on either side of the lower unit from the transom to the trim/tilt rams.

Just as the smart tower will carry a spare trailer tire, the smart boater will carry a spare prop. But stainless-steel props, which are recommended for their efficient delivery of power, are fairly expensive, so trailerboaters might want to consider carrying a prop made of composite plastic. These props are becoming more common for everyday use, and they will get you back to shore in an emergency. They're also lightweight and relatively inexpensive.

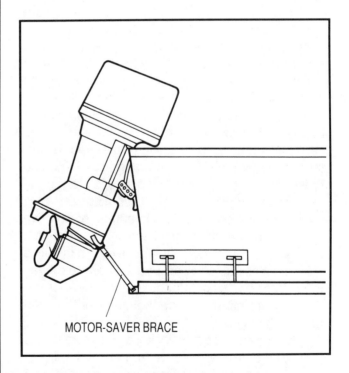

A motor-saver brace absorbs the punishment of road bumps that would normally be borne by the transom.

LOCKS AND ALARMS

The first line of defense in your battle with marine bandits is fought by an army of locks, which are now available for virtually every component of your trailer rig.

For your coupler, there's a lock that attaches to the release/lock latch. It works by preventing the lever from opening. This kind of lock can be used when the trailer is hooked to the tow vehicle, and is useful for short-term layovers. Another effective lock fastens over the hitch pin and prevents thieves from simply removing it and driving away with your hooked-up hitch platform, coupler lock and all. Another lock that prevents thieves from driving away with your trailer is one that covers the entire coupler socket from the underside.

An effective trailer-wheel locking device is a clamp or cuff that covers the hub so that the wheel can't be removed. Similar to a boot that's used by some local police departments to immobilize cars, this security measure can be created in a homemade version by running a chain or heavy-duty cable or length of pipe through the wheels, and then joining it together with a sturdy lock. Both of these locks are useful for long-term storage.

Chief among bandits' most highly prized components is the boat's valuable engine or outdrive. Transom bolt locks that fit in or over mounting bolts are the best way to foil them. Once installed, a collar spins uselessly around the bolts to prevent them from being loosened. Another kind of lock for outboard motors is one that fits over the entire mounting clamp.

A new generation of locks are also made for props, tires, trolling motors, and electronic instruments. These locks are ideal for short-term storage, but for long-term security the wiser course is to remove your valuable equipment and store it away from the trailer.

By using a simple padlock, a boater can make sure the coupler will not operate for a bandit not possessing a key.

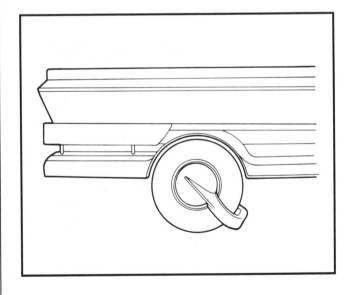

A new security accessory on the market is a wheel boot that immobilizes the trailer.

ACCESSORIES

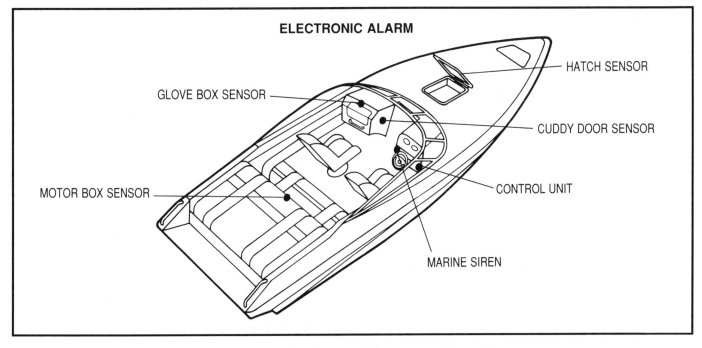

Strategically placed sensors trip an electric alarm to foil unwanted boarding.

In addition to locks, alarms are another major category of security accessories. Electronic alarms are primarily intended for boaters who keep their rigs in the water for a large part of the season.

A typical alarm system operates on either dockside power or battery power. Once activated, it prevents the engine from starting and sounds an alarm, which can be either the boat's horn or an optional siren. The alarm is triggered when a sensor connection—a magnet and a switch—is broken. Sensors are typically mounted on hatches, engine covers, doorways, outboard motors, and electronic instruments. Some units trigger strobe lights as well. Among the best are those that use no electricity until triggered and shut off after a few minutes.

A recent state-of-the-art alarm system has been introduced that uses an onboard computer. It not only empties the bilge when necessary and recharges the batteries, but also monitors sensor signals and triggers onboard security measures. These include sounding alarms, locking the engine, and phoning in a computer message to your home or the marina office. This pricey unit requires a modem, a cellular phone, and a digital communicator.

For those who wish to pursue effective but low-tech theft-prevention measures, here are a few good tips:

1. Never park a trailered boat in a dark secluded location. Make sure the site is well lit whether at home, the marina, or a motel. Remember, light is a boat thief's enemy.

2. Back up the trailer against a building, a fence, wall, tree, or other immovable object whenever possible. This restrictive placement, with the trailer remaining hooked up to the tow vehicle, will hamper a thief's ability to maneuver.
3. Remove the license plate from the trailer. If the trailer is stolen, a policeman may stop the perpetrator to check on the infraction and make inquiries.
4. If a boat is stored on land in the off-season, remove the trailer tires and set it on blocks.
5. Remove spark-plug wires if the boat is left in the water for long periods unattended.
6. Use a steering-wheel lock.
7. Remove all electronic equipment, props, water skis, and other costly accessories whenever the boat is left unattended, even overnight.

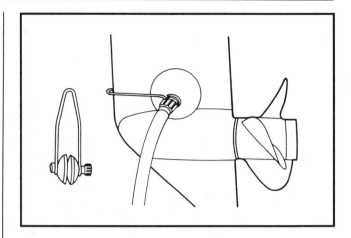

Typical motor flushers connect to a garden hose and enable freshwater to flow through the water intakes.

TOOLS

No trailerboater should embark on any trip without a basic toolbox that contains a couple of screwdrivers, a wrench set, a hammer, pliers, a knife, wire cutters, duct tape, and so forth. In addition to these tools, it's a good idea to carry a trailer jack, a complete tire-changing kit, extra fuses, and oil.

Less obvious but highly recommended traveling items include: a tow strap; a tire inflation gauge; a flat-tire repair kit; a prop wrench; a grease gun; extra rope; a flashlight; and wooden blocks to use as wheel chocks.

Other gear and tools that are important but may not be required on every trip include: motor flushers or earmuffs, which connect to the water inlets on outboards and I/Os; an oil-drain pump (electric or manual) that connects to the dipstick tube; a wire brush; touch-up paint and brushes for the trailer and engine lower unit; a hydraulic brake adjusting tool; a current tester, for troubleshooting problems with your trailer lighting system; and a tongue dolly with a tilt handle and hitch ball, which can be used to conveniently move a fully loaded trailer.

The most important thing to remember about trailerboating is that with preparation, knowledge, and the right equipment you can go anywhere and do anything.

GLOSSARY OF BOATING TERMS

Axle Ratio—the relationship between the revolutions of the drive shaft and the axle (in rear-wheel drive) or the transaxle (in front-wheel drive). The figure is expressed in a ratio, for example 3.72:1, which means that there are 3.72 revolutions of the drive shaft for each revolution of the axle.

Ball Mount Platform—the part of the hitch that the hitch ball is mounted on. Also called the hitch bar or the shank.

Beam—the dimension of a boat that's measured at the hull's widest point.

Belted Bias Tires—a type of tire that's similar to a diagonal bias tire with the addition of several plies of fabric that run through the tread area.

Bow—the forward point of the boat.

Bow Eye—a steel hook located on the bow.

Bow Stop—rubber blocks or stops located on the winch stand.

Brake Drum—the exterior housing of the brake system.

Brake Lining—the inner part of the brake drum that the brake shoes press against.

Breakaway Lanyard—an emergency safety device that triggers the brakes if the trailer becomes separated from the tow vehicle.

Bumper/Frame-Mount Hitch—a hitch that's attached to both the frame and the bumper. It may negate the 5-mph shock crash absorption feature built into the bumper.

Bumper Hitch—a hitch that's bolted solely to the rear bumper. It's not recommended for trailerboating.

Bunks—long padded or carpeted members of the trailer bed or cradle that support the weight of the boat.

Bunk Trailer—a type of trailer that uses long, padded bunks to carry the load of the boat.

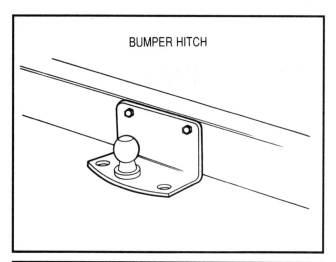
BUMPER HITCH

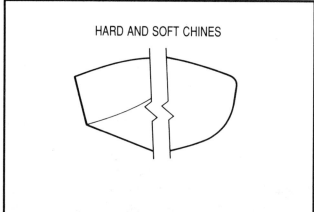
HARD AND SOFT CHINES

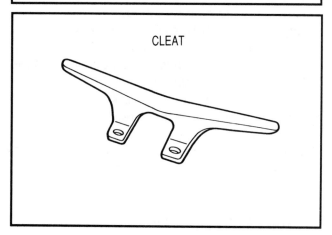
CLEAT

Chine—the point on the boat's hull where the bottom and sides meet.

Class I Hitch—a weight-carrying hitch that's designed to tow up to 2,000 lbs. gross trailer weight.

Class II Hitch—a weight-carrying or weight-distributing hitch that's designed to tow up to 3,500 lbs. gross trailer weight.

Class III Hitch—a weight-carrying or weight-distributing hitch designed to tow up to 5,000 lbs.

Class IV Hitch—a weight-carrying or weight-distributing hitch designed to tow up to 10,000 lbs.

Cleat—hardware on the deck of a boat or on a dock that's used to tie up line.

Cleating Knot—a simple crossover knot that's used to secure a boat to a cleat.

Coast Guard Safety Package—a legally required set of gear that includes personal flotation devices for each passenger on a boat, a throwable float, visual distress signals, a fire extinguisher, and a horn, whistle, or bell.

Coil Spring—a steel spring that's often used in conjunction with a shock absorber in a system called a coil-over shock. More common in cars than trailers.

Coupler—the end of the trailer tongue, which connects to the hitch ball.

Cradle—formed by bunks or rollers or a combination of both to support the trailer load.

Cruiser—a boat that has a cabin with overnight capability.

Cuddy Cabin—an enclosed living or stowage area located beneath the forward deck.

Deadrise—the V-angle of the hull usually measured at the transom.

Deep-V Hull—a wedge-shaped planing hull that has a sharp angle of deadrise.

Diagonal Bias Tires—a type of tire that's characterized by reinforcing plies or layers of

GLOSSARY OF BOATING TERMS

threaded fabric (either nylon or polyester) that crisscross through the tread area and sidewalls.

Diesel—a type of fuel used in four-cycle engines that generates tremendous amounts of compression to ignite a fuel of low volatility.

Direct Drive—a power-delivery system that tilts the inboard engine on its mountings so that the drive shaft can run in a straight line through the bottom of the hull.

Displacement Hull—a type of round-bottom hull design that smoothly displaces water while under way to reduce hydrodynamic friction.

Dolly Wheel—a part on the end of a trailer jack that gives mobility to the tongue. Instead of dolly wheels, some jacks are equipped with stationary steel feet.

Dry Weight—this figure refers to the weight of the boat prior to the filling of the fuel and freshwater tanks and the installation of optional equipment and, sometimes, the engine.

Fender—a metal or plastic cover that fits over the wheel of a trailer. Also, a cushion used to protect the hull of the boat when tied to a dock.

Fifth-Wheel Hitch—a trailer hitch that's mounted in the center of a pickup truck bed.

Fixed Ball-Mount Platform—a one-piece hitch in which the hitch ball is attached to a fixed hitch platform or bar.

Flatbed Trailer—a type of trailer that has a flat surface made of planks, plywood, steel, or aluminum that supports the trailer load.

Flat-Bottom Boat—a boat with a simple squared-off hull sometimes found on rowboats, johnboats, dinghies, and small sailing skiffs.

Flat-V—a type of planing hull that has a shallow, nearly flat deadrise.

Flushing Device—a set of clamps that goes over water inlets on either side of the lower unit of an outboard or stern-drive motor. It hooks up to a garden hose and flushes out the raw-water cooling system.

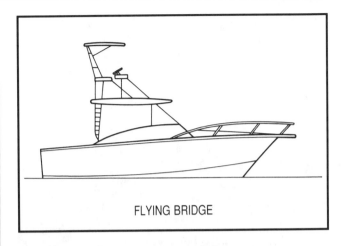

FLYING BRIDGE

Flying Bridge—a second-level helm area on a boat equipped with operational instrument controls.

Four-Cycle Motor—a gasoline or diesel engine that has a power stroke once every four cycles of the piston. Inboard and stern-drive engines are typically four-cycle motors. Also called *four-stroke*.

Frame-Mount Hitch—a hitch that's bolted or welded to the frame of the tow vehicle.

Gross Axle Weight Rating (GAWR)—the maximum allowable weight that a simple axle is designed to carry.

Gross Combined Weight Rating (GCWR)—the maximum allowable weight of the fully loaded tow vehicle and the fully loaded trailer. This figure includes all passengers and cargo.

Gross Trailer Weight Rating (GTWR)—the maximum allowable weight of the fully loaded trailer. This includes cargo. Same as GVWR.

Gross Vehicle Weight Rating (GVWR)—the maximum allowable weight of a fully loaded vehicle. This includes passengers and cargo.

Guide Bars—vertical posts mounted on the sides of the trailer to help center the boat on the cradle or bed during retrieval.

Gunwale—the upper edge of the side of a boat.

HEARST MARINE BOOKS TRAILERBOAT GUIDE

HELM

I/O Motor—stands for inboard/outboard. See *stern-drive motor*.

Keel—located at the bottom of the centerline of a sailboat or powerboat hull. Typically, it refers to a sailboat appendage that's either retractable, which means that it can slide up and out of the way, or fixed, which means that its bulb is filled with weight to act as ballast.

Keel Rollers—rubber or plastic trailer rollers that support the keel point of the hull.

Leaf Springs—a part of a vehicle's suspension system that's made of steel strips or leaves that flex when the wheels go over a bump.

Lee Side—the side of an object that's sheltered from the wind. The opposite is the windward side.

Manufacturer's Tow Package—a cluster of factory-installed trailering components that's available at the time of the tow vehicle's purchase. It can more than double the vehicle's tow rating.

Marinization—the addition of marine components to inboard and stern-drive engines, which consist primarily of automotive technology.

Helm—the driver's position or control post of a boat.

Hitch—the steel framework that attaches to a tow vehicle and carries the hitch ball.

Hitch Ball—the part of the hitch that comes in metal-to-metal contact with the coupler.

Hitch Bar—the part of the hitch that the hitch ball mounts on. Also called the ball-mount platform or shank.

Hub—a part that enables the wheel to spin around the axle.

Hull—the structural body of the boat that comes into contact with the water.

Inboard Motor—a type of gasoline or diesel engine that's located inside the hull.

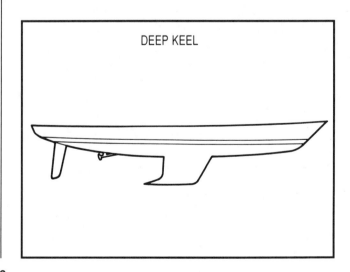

GLOSSARY OF BOATING TERMS

Mast—the vertical spar that supports a sail. It can be either stepped, which means that it can be easily taken down, or fixed, which means that it's permanently installed.

Mod-V—a type of planning hull that has a medium angle of deadrise.

Mod-VP—a type of tunnel-hull boat that has a V-bottom flanked by two sponsons. It offers the best of riding comfort and air-entrapment performance.

Outboard Motor—a marine engine that's bolted onto a boat's transom. Typical outboards are two-cycle gasoline engines. Small outboards can be portable.

Outdrive—the lower unit or drive leg of a stern-drive motor.

Overall Length—the dimension of the boat that's measured along the centerline from the transom to the bow.

Package Boat—a prerigged boat, motor, and trailer package with a long list of standard equipment that's sold with few or no options.

Personal Flotation Device—a vestlike safety jacket that aids in keeping swimmers afloat. Part of the legally required Coast Guard safety package. Also called a PFD.

Personal Watercraft—a small, lightweight craft that's similar to a motorcycle of the water. It's generally powered by a water jet motor.

Planing Hull—a type of hull that's characterized by hard chines and the ability to partially lift out of the water to reduce drag.

Pontoon Boat—a type of boat that has twin airtight, semidisplacement hulls connected above water by a platform deck. The hulls are typically made of welded aluminum.

Radial Tires—a type of tire characterized by multiple plies of fabric with the cords or thread running at a 90-degree angle from the centerline.

Receiver Box—the part of a hitch that receives and securely holds a removable ball-mount platform or hitch bar.

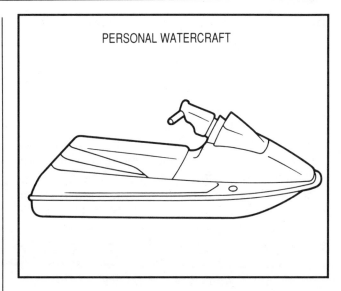

PERSONAL WATERCRAFT

Receiver Hitch—a hitch that has a removable ball-mount platform or hitch bar.

Rim—the steel part of the wheel to which the rubber tire is mounted.

Roller Trailer—a type of trailer that uses rubber or plastic rollers mounted on brackets to support the trailer load.

Runabout—a small, light recreational boat without overnight amenities.

Safety Chains—chains that connect the trailer tongue to the tow vehicle. They secure the trailer to the tow vehicle in case the coupler detaches.

Semidisplacement Hull—a round-bottom hull design with soft chines. It has some planing characteristics.

Shank—the part of the hitch that the hitch ball mounts on. Also called the hitch bar or hitch shank.

Spindle—the part of the axle that the wheel bearings rotate around.

Sponsons—the water contact points on a tunnel-hull boat.

Spring Line—a line that's cleated amidships on the boat or dock.

Step Bumper Hitch—a hitch that's built into the rear bumper, typically on a pickup truck. It usually has attachment points on the vehicle frame, and if it doesn't, it's not recommended for trailerboating.

Stern-Drive Motor—a type of marine engine that combines the best of the outboard and inboard worlds. The engine is mounted inboard and the drive shaft runs through a cutout in the transom to an exposed lower unit or drive leg. Most stern drives use gasoline engines. Also called I/Os.

Surge Brakes—a hydraulic trailer braking system that's activated by a sudden slowing of forward momentum when the tow-vehicle brakes are applied.

Sway—a side-to-side wandering movement of the trailer behind the tow vehicle while under way.

Sway Control Device—attaches to the trailer tongue and the tow vehicle and uses friction to resist pivoting or sway movement of the trailer while driving. It is generally a quick fix that may temporarily mask the effects of a larger problem.

Tandem-Axle Trailer—a trailer that has two axles for extra strength and stability. Also called a dual-axle trailer.

Tiller-Handle Motor—a type of motor that requires the boater to use a control arm for steering while under way.

Tilt-Bed Trailer—a type of trailer that has a hinged tongue or hinged frame, which enables the bed to tilt like that of a dump truck.

Tongue—see *Trailer Tongue*.

Tongue Jack—a jack mounted on the trailer tongue that raises the coupler to connect with the hitch ball.

Tongue Weight—the weight of the loaded trailer

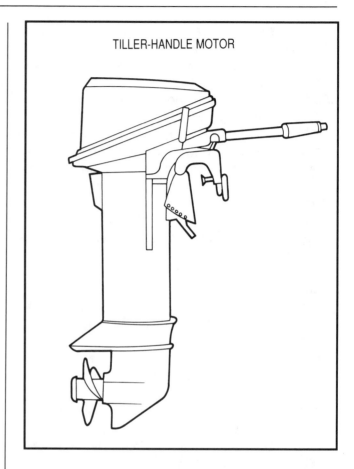

TILLER-HANDLE MOTOR

on the hitch. It should be a maximum of 10 percent of the trailer load for most applications.

Torsion Bar—a trailer suspension system that has a hexagonal exterior axle made of tubular steel that encloses a three-sided solid-steel shaft, which in turn is surrounded by rubber inserts. The rubber inserts absorb the twisting motion caused by the wheels going over bumps.

Tow Rating—the maximum amount of weight a vehicle is rated to tow.

Trailer Bed—formed by a pattern of bunks or rollers or a combination of both to support the trailer load.

GLOSSARY OF BOATING TERMS

Trailerboat—a marine craft that's generally no more than twenty-six feet long and eight feet six inches wide.

Trailering Height—the measurement between the ground and the bottom of the hull when it's mounted on a trailer plus the height of the boat.

Trailer Tongue—the forward end of the trailer nearest the tow vehicle. The coupler and other components are mounted on the tongue.

Tri-Hull—a type of planing hull that's characterized by three V-shaped, side-by-side bottom components. Also called a cathedral hull.

Trim—a boat operation term that refers to the running position of the engine's drive unit. On boats that are equipped with power trim, the drive unit can be raised or lowered by pressing a button.

Tunnel-Hull Boat—a type of high-performance boat characterized by twin hulls that trap air and create lift for minimum drag.

Two-Cycle—gasoline engine technology in which a power stroke is delivered every other cycle of the piston. Outboards are typically two-cycle engines. Also called two-stroke.

V-Drive—a power-delivery system that splits the drive shaft to form a V-angle before running the prop shaft through the bottom of the hull.

V-Hull—a wedge-shaped planing hull.

Weight-Distributing Hitch—a frame-mounted hitch that uses spring bars to apply leverage between the tow vehicle and the trailer to distribute tongue weight to all wheels of the tow vehicle.

Wheel Bearings—two rings of steel rollers inside the hub that enable the wheel to rotate freely around the spindle.

Wheel-Bearing Protectors—caps that fit over the wheel hubs to protect the wheel bearings by keeping grease under pressure to prevent water from getting in.

Winch—a manual or electric mechanism used to haul or hoist a boat.

Wiring Harness—the heavy-duty plug that connects the tow-vehicle and trailer electrical systems.

INDEX

Page numbers in *italics* refer to illustrations.

A-arm suspensions, 37, *38*
accessories, 83–92
　alarms, 91
　anchors, 83
　boat covers, 80, *80, 83, 83*
　Coast Guard safety packages, 47, 83, *84,* 94, 97
　double-lip dust caps, 87
　hitch-ball covers, 88
　locks, 90
　marine VHF radios, 83
　mirrors, extended, 88
　motor-saver braces, 89, *89*
　mud flaps, 88
　multiball hitch platforms, 89
　PFD (personal flotation devices), 83, *84,* 97
　power winches, 88
　props, spare, 89
　ratchet-buckle tie-down straps, 88, *88*
　tongue jacks, *14, 22,* 23, 86, 98
　tongue walk ramps, 88
　tools, 92
　trailer guides, 88
　trailer jacks, 86, *86,* 92
　trailer tire spare mounts, *87*
　water-sealed submersible lights, 87
　wheel-bearing protectors, 86, 99
adjustable ball mounts, *29*
air-bag suspension, 37, *38*
alarms, 91, *91*
all wheel drive (AWD), 41
anchor lockers, *4*

anchors, 83
automotive engine blocks, marinized, 9
automatic transmissions, 41, 43
auxiliary transmission coolers, 41
AWD (all wheel drive), 41
axle ratio, 42, 93
axles, 19–20, *19*
　drop, 19
　Gross Weight Rating, 18, 37, 95
　ratio, 39, 42, 93
　shafts of differential, *39*
　single configuration, 19, 24
　spindles, *19,* 98
　tandem configuration, 19–20, 24, 98
　tandem roller trailers, *15*
　tow vehicles, rear, *40*
　transaxles, 39–40
axle shafts of differential, *39*

backing up, 49–52, *50, 51*
balls, hitch, *26,* 30, *30,* 75
　definition, 96
　solid-steel, 34
　winterizing, 78
ball mounts, *26,* 29
ball-mount platforms (hitch bars, shanks), *26,* 30, 34, 93, 96, 97
basic driving techniques, 47–49, *48*
bass boats, *4*
beams, 6, 93
bearings, wheel, 19
　definition, 99
　maintenance, 72, *73*

101

INDEX

belted bias tires, 17–18, *17,* 93
blind spots, driving, 52
boats:
　accessories, 83–86
　bass, *4*
　center console fishing, *4*
　checkup, 47
　covering, 80, *80,* 83, *83*
　deep keel, 12
　dock maneuvering, 61–66
　fixed mast, 12
　handling, 66–67
　hauling out, 68–69
　hull shapes, 6–8
　hull specifications, 5–6
　launching, 57–60
　maintenance, 75–76
　matching trailers and, 23–24
　package, 13, 97
　pontoon, *5,* 8, 97
　power sources, 8–12
　tournament ski, *3*
Boat/U.S. Foundation, 66
bow eyes, *2,* 93
bow hatches, *3*
bows, *2,* 93
bow seating, *3*
bow stops, *14, 22,* 23, 93
　maintenance, 75
boxes, receiver, 26, 97
braces, motor-saver, 89, *89*
brake drums, 93
brake linings, 93
brakes, 20
　adjusting slots, *73*
　adjusting tools, *73*
　bleeding (for winterization), 78, *78*
　electronic, 20
　maintenance, 73–74, *73*

brake shoes, winterization, 77
braking fluid, maintenance, 75
breakaway lanyards, 93
bumper/frame-mount hitches, 28, 34, 93
bumper hitches, 28, *28,* 93, *94*
bunk trailers, 14–15, *14,* 93
　PWC, *15,* 23
buying, *see* choosing

capacity:
　carrying, 24
　tow vehicle, 35–37
carburetor intakes, 80
carrying capacity, 24
cars, *see* tow vehicles
cathedral hulls (tri-hulls), 6, *7,* 99
cavitation plates, *9,* 76, 78
center console fishing boats, *4*
chains, safety, *14, 22, 26, 32, 32, 69,* 75, 97
Chapman Piloting: Seamanship & Small Boat Handling (Maloney), 66
checkups, *see* safety checks
chine, hard and soft, 6, 23, 94, *94*
chocks, 81, 92
choosing:
　hitches, 34
　tow vehicles, 42
　trailers, 23–24
CID (cubic-inch displacement), 37

classification of hitches, 25–27, 94
　tongue weight, 25, 38, 40, 42, 46, 98
　see also Gross Vehicle Weight Rating; *specific classifications*
Class I hitches, 26, *27,* 94
　ball capacity, 30
　matching hitches to trailers, 34
　safety chain capacity, 32
Class II hitches, 26, *27,* 94
　ball capacity, 30
　matching hitches to trailers, 34
　safety chain capacity, 32
Class III hitches, *27, 27,* 94
　ball capacity, 30
　matching hitches to trailers, 34
　safety chain capacity, 32
Class IV hitches, *27, 27,* 94
　ball capacity, 30
　matching hitches to trailers, 34
　safety chain capacity, 32
cleats, *2,* 94, *94*
　knots, 94
　lines, 61, *61*
Coast Guard, U.S., 66
Coast Guard safety package, 47, 83, *84,* 94
　PFD (personal flotation devices), 83, *84,* 97
cockpits, *2, 4*
cog wheels, *73*
coil-over shock suspensions, 21, 36, 38, *38,* 94

coil springs, 21, 38, 94
　maintenance, 74
consoles:
　side, *2*
　twin, *3*
control arms, coil-over shock suspensions, *38*
cooling systems, 41
couplers, *14, 22, 26,* 31, 94
　locks, 90, *90*
　maintenance, 75
　rust, 74
covers:
　boat, 80, *80,* 83, *83*
　engine, *3*
　hitch-ball, 88
cradles, 94
cranes, *60*
cruisers, 94
cubic-inch displacement (CID), 37
cuddy:
　cabin, 94
　door sensor, *91*
curb weight vs. tow rating, 37

deadrise, 6, 94
decks, pontoon boats, *5*
deck soles, *3*
deep keel, *96*
　sailboats, 12
deep-V hulls, *7,* 94
diagonal bias tires, 16–18, *17,* 94
diesels, 95
　inboard engines, 9–10
differentials, *39, 40*
housing, *39*
dihedrals, 8
direct drives, 10, 95
displacement hulls, 95
docks:
　maneuvering, 61–65, *62, 63, 64, 65*

INDEX

pulling away from, 64, *64, 65*
pulling into, 62, *63*
dolly wheels, *22,* 86, 95
double-lip dust caps, 87
drive legs (outdrive units), 11, *11,* 97
drivelines, tow vehicles, 40
drive shafts, *40*
   inboard engines, *10*
   packing, 79
drive systems, 39–41
driving techniques:
   backing up, 49–51
   basic, 47–49
   defensive, 52
   trailer sway, 52–55, *53,* 98
drop axles, 19
dry weight, 6, 95
dual-axle (tandem-axle) roller trailers, *15,* 19–20, 24, 98
dual outboard engines, *4*

electric trolling motors, *4*
electric winches, 23
electronic alarms, *91*
electronic brakes, 20
engine power, tow vehicles, 37
engines:
   accessories, 89
   changing oil of, 79–80
   coupler splines, 79
   covers, *3*
   diesel inboard, 9–10
   drive shafts, 10
   dual outboard, *4*
   flushing out, *68,* 92, 95
   four-cycle, 8–9, 10–11, *10, 11,* 95, 98
   inboard, 9–10, 96
   inboard/outboard (I/O, stern-drive), 10–11, *11,* 79, 98
   maintenance, 76, 77

marinized automotive, 9
oil coolers, 41
outboard, *4, 5,* 8–9, 79, 97
outdrive units (drive legs), 11, *11,* 97
packing seals of inboard, *10*
prop shafts, *10*
tow vehicles, 37, *40*
trimming, *66, 67, 69*
two-cycle, 8, 9–10, *9,* 99
equalizer bars, maintenance, 74
equalizing (weight-distributing) hitches, 29–30, *29,* 34
extended mirrors, 88
extension tongues, 22

fenders, *14,* 95
fifth-wheel hitches, 29, *29,* 95
figure-eight cleating knots, 61, *61*
fishwells, *4*
fixed ball-mount platform hitches, 28, *29,* 95
fixed-mast sailboats, 12
flame arrestors, 79
flaps, mud, 88
flatbed trailers, 16, *16,* 23, 95
flat-bottom hulls, *7,* 8, 95
flat-tires, 86
   repair kits, 92
flat-V hulls, 95
flotation devices, personal (PFD), 83, *84,* 97
fluid reservoirs, surge brakes, *20*
flusher couplings, 76, *77*
flushing devices, 92, 95
flying bridges, 95, *95*

foredecks:
   cuddy cabin, *3,* 94
   pontoon boats, *5*
forward components, 22–23, *22*
   bow stops, *22,* 23, 75
   tongues, *14,* 22, *22,* 75, 98
   winches, 22–23, *22, 69,* 75, 99
four-cycle engines:
   definition, 95
   inboard, *10*
   outboard, 8–9
   stern-drive (I/O), 10–11, *11,* 98
four-wheel drive (4WD), 39, 41, 43
frame-mount hitches, 28–29, 34, 95
frames, maintenance, 74
front end, trailers, 22–23, *22*
front-wheel drive (FWD), 39–41, 42–43
fuel filters, 79
fuel-line screens, 79
FWD (front-wheel drive), 39–41, 42–43

GAWR (Gross Axle Weight Rating), 18, 37, 95
GCWR (Gross Combined Weight Rating), 36–37, 42, 95
gearcases, outboard motors, *9*
gimbal bearings, 79
glossary of boating terms, 93–97
glove box sensors, *91*
grease guns, 92
Gross Axle Weight Rating (GAWR), 18, 37, 95

Gross Combined Weight Rating (GCWR), 36–37, 42, 95
Gross Trailer Weight Rating (GTWR), 95
Gross Vehicle Weight Rating (GVWR), 20
   definition, 95, *95*
   hitch capacity, 25
   matching boats to trailers, 24
   matching hitches to trailers, 34
   minimum breaking strength for chains, 32
GTWR (Gross Trailer Weight Rating), 95
guide bars, 95
guides, trailer, 88
gunwales, *2,* 95
GVWR, *see* Gross Vehicle Weight Rating

handling, boats, 66–67, *66, 67*
hand position, backing up, *50*
hand-wheel-type couplers, 31, *31,* 34
hangers, maintenance, 74
harnesses, wiring, *26, 33, 33,* 99
hatch sensors, *91*
hauling out, 68–69, *69*
height, trailering, 6, 99
helms, *2,* 96, *96*
   pontoon boats, *5*
hinged tongues, 22–23
hinge pins, 79
hitches, 25–34
   ball-mount platforms (shanks), *26,* 30, 34, 93, 96, 97

103

INDEX

hitches *(cont.)*
　balls, *26,* 30, *30,* 34, 75, 78, 96
　bumper/frame-mounts, 28, 34, 93
　bumpers, 28, *28,* 93, *94*
　Class I, 26, *27,* 30, 32, 34, 94
　Class II, 26, *27,* 30, 32, 34, 94
　Class III, 27, *27,* 30, 32, 34, 94
　Class IV, 27, *27,* 30, 32, 34, 94
　classifications, 25–27, 94
　connections, *26*
　couplers, *14, 22, 26,* 31, 74, 75, 90, *90,* 94
　definition, 25, 95
　fifth-wheel, 29, 95
　fixed ball-mount platforms, 28, 29, 95
　frame-mount, 28–29, 34, 95
　load-equalizing (weight-distributing) hitches, 29–30, *29,* 34
　maintenance, 75
　matching to trailers, 34
　receiver hitches, 28–29, 97
　receiver-type hitches, 34
　safety chains, *14, 22, 26,* 32, *69,* 75, 97
　surge-brake cables, 33
　types, 28–29
　wiring harnesses, *26,* 33, *33,* 99
　see also specific types of hitches
hoists, *60*

hubs, 96
hulls:
　definition, 96
　maintenance, 75–76
　matching boat to trailer, 23
　primary specification, 5–6
　shape, 6–8
　see also specific hull shapes
hydraulic brakes, 20
　bleeding (for winterization), 78, *78*

identification numbers, *83*
inboard engines, 9–10, *10,* 96
inboard/outboard (I/O, stern-drive) engines, 10–11, *11,* 98
　storage, 79
inflation, tires:
　gauges, 92
　pressure, 71–72
inside outboard motors, 9

jacks:
　tongue, *14, 22,* 23, 86, 98
　trailer, 86, *86,* 92

keel, 11–12, 96
　rollers, 15–16, 96
knots, cleating, 61, *61*

launching, 57–60
　prelaunch preparation, 57–58, *58*
　procedure, 59–60, *59, 60*
　ramp inspection, 58
　by sling/lift, 60, *60*
leaf clips, maintenance, 74

leaf springs, 21, *36,* 37, *38,* 96
　maintenance, 74
leaning posts, *4*
lee side, 96
length, overall, 5–6, 97
lever-type couplers, 31, *31,* 34
lifts:
　hauling out by, 68
　launching by, 60
lights, 22
　maintenance, 74–75
　water-sealed submersible, 87
limited-slip differentials, 41
livewells, *4*
load-equalizing (weight-distributing) hitches, 29–30, *29,* 34
load rating, tires, 18
locks, 90
lower units, outboard motor, *9*

maintenance, 71–81
　boats, 75–76
　brakes, 73–74
　engines, 76, 77
　frames, 74
　hitches, 75
　lights, 74–75
　suspension systems, 74
　tires, 71–72
　tongue components, 75
　wheel bearings, 72, *73*
　wheels, 72
　winterizing, 77–81
　wiring, 74–75
Maloney, Elbert S., 66
maneuvering, dock, 61–65, *62, 63, 64, 65*
manual brake adjustment, 77

manual hand-crank winches, 23
manual transmissions, 41, 43
manufacturer's tow packages, 42, 43, 96
marine power, launching, 59–60, *59*
marine sanitation devices (MSD), 76, 78
marine sirens, *91*
marine VHF radios, 83
marinization, 96
　of automotive engine blocks, 9
master cylinders, surge brakes, *20*
masts, 11–12, 97
matching:
　boats to trailers, 23–24
　hitches to trailers, 34
maximum capacity rating, 71
minivans, *see* tow vehicles
mirrors, extended, 88
mod-V hulls, *7,* 97
mod-VP, 8, 97
motor box sensors, *91*
motor flushers, 92
motors, *see* engines
motor-saver braces, 89, *89*
mounts, trailer tire spare, *87*
MSD (marine sanitation devices), 76, 78
mud flaps, 88
multiball hitch platforms, 89

observer seats, *3*
oil, changing, 79–80
oil-drain pumps, 92
oil filters, 79
outboard engines:
　bass boats, *4*

INDEX

definition, 97
dual, *4*
four-cycle, 8–9
pontoon boats, *5*
storage, 79
outdrive units (drive legs), 11, *11,* 97
overall length, 5–6, 97
overload springs, 37

package boats, 13, 97
packing seals, inboard engines, 10
passenger cars, *see* tow vehicles
pedestal seat mounts, *4*
personal watercraft, 97, *97*
PFD (personal flotation devices), 83, *84,* 97
pinion gears, *39,* 79
piston/actuators, surge brakes, 20, *20*
planing hulls, 6, 97
pneumatic lines, air-bag suspensions, *38*
pontoons, *5*
  boats, *5,* 8, 97
  boat trailers, *16*
power sources, 8–12
  effect on trailering, 11–12
  inboard engines, 9–10, 96
  inside/outboard motors, 9
  outboard motors, 4, 5, 8–9, *9,* 79, 97
  stern-drive engines (I/O), 10–11, *11,* 79, 98
  tow vehicle engines, 37, 40
power winches, 88
prelaunch preparation, 57–58, *58*

props, *9,* 76, 78–79, *79*
  spare, 89
  trimming, 66–67, *66, 67*
prop shafts, inboard engines, 10
prop wrenches, 92
pulling away, docks, 64, *64, 65*
pulling into docks/slips, 62, *63*
purchasing, *see* choosing
PWC bunk trailers, *15,* 23

radial tires, 17–18, *17,* 97
radiators, 41
radios, marine VHF, 83
raised casting decks, *4*
ramp inspection, 58
ratings:
  Gross Axle Weight, 18, 37, 95
  Gross Combined Weight, 36–37, 42, 95
  Gross Trailer Weight, 95
  maximum capacity, 71
  tire load, 18
  tongue weight, 42
  tow, 35–37, 99
  see also Gross Vehicle Weight Rating
rear axles, tow vehicles, *40*
rear suspension, 42
rear-wheel drive (RWD), 39–41, 42–43
receiver boxes, *26,* 97
receiver hitches, 28–29, 34, 97
registration, *83*
removable ball-mount platform hitches (hitch bars), 30, 34
removable tongues, 22
retrieving boats, *see* launching

reversible ball mounts, 29
rims, 97
ring gears, *39*
rod lockers, *4*
rollers, *14*
roller trailers, 15, 23
  definition, 97
  tandem-axle, *15*
round-bottom hulls, *7,* 8
rudders, 79
runabouts, 97
rust:
  brake shoes, 77
  couplers, 74
  frames, 74
  wheels, 72, 77
RWD (rear-wheel drive), 39–41, 42–43

safety chains, hitches, *14, 22, 26,* 32, *32, 69, 75,* 97
safety checks:
  boats, 47
  on-the-road, 55
  tow vehicles, 45–46
  trailers, 46–47
safety latches, *69*
safety packages, Coast Guard, 47, 83, *84,* 94
  PFD (personal flotation devices), 83, *84,* 97
sailboats, fixed-mast, 12
screw-type couplers, 31, *31*
semidisplacement hulls, 8, 97
sensors, security, 91
shackles, maintenance, 74
shanks (ball-mount platforms, hitch bars), *26,* 30, 34, 93, 96, 97
shock absorbers, *36,* 37, 38

shock/pistons, *20*
side consoles, 2, *4*
single-axle configuration, 19, 24
sirens, marine, *91*
skegs, *9,* 76, 78
ski boats, tournament, *3*
ski pylons, *3*
ski tow eyes, *2*
slings:
  hauling out by, 68
  launching by, 60, *60*
spare mounts, trailer tires, 87
spark plugs, 80
specifications, understanding tire, *18*
spindles, 19, *19,* 98
splashboards, *4*
sponsons, 8, 98
sport utilities, *see* tow vehicles
spring lines, 98
stabilizer bars, 37, *38*
Stapleton, Sid, 66
Stapleton's Powerboat Bible (Stapleton), 66
steering-cable rams, 79
step-bumper hitches, 28, 34, 98
stern-drive engines (I/O), 10–11, *11,* 98
storage, 79
stern steering, *62*
storage, long term, 77–81
straight axles, 19
sun lounger seats, *3*
surge brakes, 20, *20*
  cables, 33
  definition, 98
  fluid reservoirs, *20*
  housing, *14, 22*
  maintenance, 73–74, *73,* 75

105

INDEX

surge brakes *(cont.)*
   master cylinders, *20*
   piston/activators, *20*
suspensions:
   maintenance, 74
   systems, 21, *21*
   tow vehicles, 37–38
sway:
   control devices, 98
   trailers, 52–55, *53*, 98
swim platforms, *2*
swivel pins, 79

tandem-axle (dual-axle) roller trailers, *15*, 19–20, 24, 98
theft-prevention measures, 91–92
thermostatic clutches, 41
throttles, *2*
tiller-handle motors, 8, 98, *98*
tilt-bed trailers, 98
tires, 17–18
   accessories, 92
   belted bias, 17, *17*, 93
   correct load rating, 18
   diagonal bias, 17–18, *17*, 94
   flat, 86, 92
   inflation gauges, 92
   inflation pressure, 71–72
   maintenance, 71–72
   maximum capacity rating, 71
   radial, 17, *17*, 97
   specifications, *18*
   trailer spare mounts, 87
   winterization, 77
tongues, *14, 22, 22*
   definition, 98
   dollies, 92
   hitch connections, *26*
   jacks, *14, 22*, 23, 86, 98
   maintenance, 75

weight, 25, 38, 40, 42, 46, 98
weight rating, 42
tongue walk ramps, 88
toolboxes, recommended, 92
torque, 37, 39, 42
torsion bar suspension systems, 21, *21*, 38, 98
   maintenance, 74
tournament bass boats, *4*
tournament ski boats, *3*
tow rating, 35–37, 99
tow straps, 92
tow vehicles, 35–43
   accessories, 88–89
   checkup, 45–46
   choosing, 42
   cooling systems, 41
   drive systems, 39–41
   engine power, 37, 40
   Gross Combined Weight Rating, 36–37, 42, 95
   hauling out, 68–69
   launch, 57–60
   manufacturer's packages, 42, 43
   suspensions, 37–38
   towing capacity, 35–37
   transmissions, 41
   see also Gross Vehicle Weight Rating; tires
trailerboats, *see* boats
trailering height, 6, 99
trailers, 13–24
   accessories, 86–88
   axles, *see* axles
   beds, 99
   brakes, 20, 73–74, *73*, 78
   bunks, 14–15, 23, 93
   checkup, 46–47
   flatbed, 16, *16*, 23, 95
   forward components, 22–23

Gross Combined Weight Rating, 36–37, 42, 95
Gross Trailer Weight Rating, 95
guides, 88
hauling out, 68–69
keel-roller, 15–16, 96
launch, 57–60
lights, 21–22, 74–75, 87
matching boats and, 23–24
roller, 15, 23, 97
spare tire mounts, *87*
suspension systems, 21, *21*, 74
sway, driving, 52–55, *53*, 98
tires, *see* tires
types, 14–17
wheel bearings, 19, 72, *73*, 99
wheel locking devices, 90
see also specific types of trailers
trailing arms, 37, *38*
transaxle, 39–40
transmissions, *36, 40*, 41, 43
   auxiliary coolers, 41
transoms, *2, 9*
   bolt locks, 90
   drain plugs, replacing, 58
   lifting hooks, *2*
tri-hulls (cathedral hulls), 6, *7*, 99
trim, 8, 99
trimming engines, *69*
   effects of, *66, 67*
trim/tilt assemblies, 79
trucks, *see* tow vehicles
tunnel-hulls, *7*, 8
   definition, 99
   sponsons, 8, 98
turning radius, *48*

twin consoles, *3*
two-cycle engines, 8, 9–10, *9*, 99
two-wheel drives (2WD), 39, 41, 43

U-joints, 79
United States Power Squadrons, 66
universal joints, tow vehicles, *40*

V-drive, 10, 99
vehicles, *see* tow vehicles
vent plugs, 79
vertical rod holders, *4*
VHF radios, marine, 83
V-hulls, 6, *7*, 8, 99

watercraft, personal, 97, *97*
water pumps, high-performance, 41
weight, trailers, 18
weight-distributing (equalizing) hitches, 29–30, *29*, 34
weight ratings, 54, *54*
   Gross Axle Weight Rating, 18, 37, 95
   Gross Combined Weight Rating, 36–37, 42, 95
   Gross Trailer Weight Rating, 95
   see also Gross Vehicle Weight Rating
wheel-bearing protectors, 19, *19*, 72, *73*, 86
   definition, 99
   winterization, 77
wheel bearings, 19
   definition, 99
   maintenance, 72, *73*
wheel boots, 90, *90*

INDEX

wheels:
   maintenance, 72
   trailer locking devices, 90, *90*
   winterization, 77

width, hulls, 6
winches, *14,* 22–23, *22, 69*
   cables, *22*
   definition, 99
   hooks, *14*

lines, maintenance, 75
power, 88
stands, *14*
windscreens, *2*
winterization, 77–81

wire brushes, 92
wiring:
   harnesses, *26,* 33, *33,* 99
   maintenance of, 74–75